T0174454

CAMBRIDGE COUNTY GEOGRAPHIES

General Editor: F. H. H. GUILLEMARD, M.A., M.D.

DURHAM

Cambridge County Geographies

DURHAM

by

W. J. WESTON, M.A., B.Sc.

With Maps, Diagrams and Illustrations

Cambridge:

at the University Press

1914

CAMBRIDGE UNIVERSITY PRESS
Cambridge, New York, Melbourne, Madrid, Cape Town,
Singapore, São Paulo, Delhi, Mexico City

Cambridge University Press
The Edinburgh Building, Cambridge CB2 8RU, UK

Published in the United States of America by Cambridge University Press, New York

www.cambridge.org
Information on this title: www.cambridge.org/9781107694651

First published 1914
First paperback edition 2013

A catalogue record for this publication is available from the British Library

ISBN 978-1-107-69465-1 Paperback

CONTENTS

ILLUSTRATIONS

MAPS AND DIAGRAMS

The illustrations on pp. 8, 17, 20, 24, 28, 49, 50, 52, 56, 57, 96, 111, 119, 129, 131, 137, 139, 156, 157, and 171 are reproduced from photographs by Messrs J. Valentine & Sons; those on pp. 42, 46, 107, 117, 121, 125, 133, 136, 140, 143 and 149 from photographs by Mr Godfrey Hastings; those on pp. 7, 12, 22 and 150 from photographs by Mr A. H. Robinson; those on pp. 76, 78 and 80 were kindly supplied by Messrs Palmers Shipbuilding Co. Ltd., from photographs by Messrs W. Parry & Son; that on p. 3 is from a photograph by Mr A. Bailes; that on p. 163 is from a photograph supplied by Messrs Mansell & Co.; those on pp. 87 (from a photograph by Messrs Illingworth & Co.), 92, 98 (from a photograph by Messrs W. S. Rickers), 115 (from a photograph by Mr Fulthorpe) and 152 (from a photograph by Mr A. Moffat) are from the *North Eastern Railway Magazine*.

1. County and Shire. The Word Durham: its Origin and Meaning.

When the French-speaking Norman kings made plans for the firm ruling of England—for the keeping of peace and order, and the raising of money to pay the expenses of government—they called the divisions of the country by the name to which they had grown accustomed. So all the divisions are still called counties—districts ruled by a *Comte* or Count: thus we speak of the county of Northumberland as well as of the county of Durham.

Probably the Normans made few new divisions. They appointed governors of the divisions they found, and these divisions already had names. Some were once separate kingdoms, like Essex, the kingdom of the East Saxons. Others were parts *shorn* off—*shares* or *shires* (from an Old English word, *scir*, *sciran*, to cut, to divide)—from the ancient kingdoms, like Hampshire, which was part of the old kingdom of Wessex. The English name would remain and the Norman one be added; so that, while all the divisions are counties, some are also shires.

The word shire does not, however, always mean quite the same as county: it is still sometimes used in its old

sense of a part.   Richmondshire, for example, is the part
of Yorkshire round Richmond; and Hallamshire is the
part round Sheffield.   Most interesting, however, for us is
Norhamshire (North-durham-shire)—that part of Northumberland along Tweed bank, till the middle of the
nineteenth century a part of Durham county.

Durham, however, has from the earliest times been
looked upon as a district apart, not simply as a piece
shorn off from a larger division.   It has never, that is,
been called a *shire* as its big neighbour on the south has.
The old kingdom of Northumbria, "the land north of
the Humber," included Durham and Yorkshire as well
as Northumberland.   But the space between Tyne and
Tees was regarded as a close dependency of the Church
and its governor was not a king but a bishop.   The
whole county formed the land of the cathedral on the
Wear.   It was the "patrimony of St Cuthbert," the
dominion ruled over by the successors of this great
missionary.   Not till late in the Middle Ages did men
speak of it as County Durham.   In the survey that
William the Conqueror ordered to be taken and recorded
in Domesday Book it is called Saint Cuthbert's Land.

It was Saint Cuthbert of Lindisfarne who chose the
great grey rock almost surrounded by the Wear as his
last resting-place.   No better choice in those early days
of war and pillage could have been made.   Dunholm or,
as it afterwards became, Durham, is the finest example
we have of a hill and river fortress.   A great rock rises
almost straight up from the water, which curves round it
in a loop that very nearly makes the rock (*dun*) into an

island (*holm*). Defence against enemies, a plentiful supply of water, and a healthy position were here found together.

In later times Durham, because of its position near the northern border and far away from the king's court,

St Cuthbert's Cross

was a County Palatine. Its ruler, the earl-bishop, dwelt in a palace with his retinue, or travelled from one to another of his many manors in order to judge or punish

or raise money. London was then so distant and in-
accessible that the bishop acted in most ways as a king.
His palace was the castle that grew up in the shadow of
the cathedral; and he himself was at times more a warrior
than a priest. To him was entrusted the important duty
of defending the northern border, and even the great
church was

> Half church of God, half castle 'gainst the Scots.

It was at Neville's Cross, not a mile from the cathedral,
that King David of Scotland was defeated and captured
in 1346; and at all times the bishop had to be ready to
repel inroads or to march with his forces northwards.
Durham is still called a County Palatine, though its
bishop is not now a temporal prince but occupies himself
in church affairs alone. Even now, as one memorial
of his former importance, he shares with the bishops of
Winchester, the ancient capital, and London, the modern
one, the distinction of being always a member of the
House of Lords.

2. General Characteristics.

Durham in these days means to most of us collieries
and the ironworks dependent on them; for it is the rich
coal deposit along the flanks of the Pennines and in the
coast plain that provides power for the many industries
crowded into the county. The ports on the south bank
of the Tyne, from Dunston twelve miles up river to
South Shields right at the mouth, are coaling ships day

and night. Iron ore, and lime to carry off its impurities, are here in abundance, so that Durham rivals Staffordshire as a seat of the iron and steel industry. It is a typical industrial region. And fringing the coast from Shields to Port Clarence are ports, the homes of sailors and engineers and fishermen. In the oblong between the sea on the east and a line joining Barnard Castle to Shotley Bridge on the west, the people are crowded so closely together that three counties only are more densely populated. As we shall see later, Lancashire, Middlesex, and Glamorgan alone have more people to the square mile.

No one would guess Durham to be an "industrial" county who only traversed the district west of this line— a district first of fertile fields and then of moorland pasture, a district that as we go westward becomes more and more destitute of traces of the activity of man. At the meeting-place of Cumberland, Westmorland, and Durham, the broad divide that like a great uneven slab of stone lies along the middle of northern England raises itself half a mile above the surface of the sea. The wide moor with its flattened and rounded top worn down with the weather of ages is called Cross Fell. It is the highest part of the Pennines. We can hardly speak of it as the highest *point*; for on these uplands the limestone and millstone grit weathers into rounded outlines, not into peaks, and we may walk for miles without having to scale any rugged rocks or scramble down a precipice.

From Cross Fell the dales of the Derwent, the Wear, and the Tees slope eastward towards the North Sea, narrow at first, but as they reach the plain widening like three folders of a fan. The deep dales with their steep sides merge into the level along the coast ; the enclosing hills leave greater space and lessen in height; and obstacles in the path of the hurrying waters become fewer. Instead of leaping over these obstacles or of forcing a way through them, the rivers now curve gently aside. Their slow tranquil currents are now better fitted for the service of man. He uses them as roads, sets up his workshops along their banks, and builds here the steel vessels for which the Tyne and Wear are famous.

In the uplands we note at rare intervals stone quarries, like the limestone ones that provided the " Frosterley marble " of which the wonderful pillars in the cathedral are made. Or we may chance on a deserted lead-mine, or more seldom on one still working. For Durham is no longer what it once was, the chief lead-producing district of Europe. Imported ore from Australia or Chile is so cheap that lead-mining is here a rapidly-declining industry.

Other marks of man's activity are few indeed ; we may wander many an hour without seeing a farmstead or even a shepherd's hut. But when we near the lower land, silence gives place to the dull, far-off roar of the distant factories, houses become more numerous and cluster into villages and towns, till as we approach the sea we pass through one of the most thickly peopled parts of Britain or indeed of any part of the world.

As we stand on the slope of Burnhope Seat and look eastward across the lofty moors covered with heather and peat, their rounded outlines looking rather dreary and

The City of Durham

desolate, we may roughly picture to ourselves the small compact county through which we travel in this book. Right below us is Wearhead, and for some distance we trace the winding way of the river among the hills. With imagination for our helper we follow the infant stream carving a path for itself, past the sheep-runs, the quarries of hard sandstone, and the lead-mines of its upper

Caldron Snout

course, to where Durham, a city built on a hill, dominates the landscape round. On a clear Sunday, when no smoke from the factories darkens the air, its stately cathedral is visible even from the sister capital, Newcastle, sixteen miles away.

Below Durham the river winds in a slower, calmer current towards that other place of ancient memories, Monkwearmouth, whose church is even older than the cathedral itself or than that of Bede at Jarrow. But here the past is "like an unsubstantial pageant faded," overwhelmed by the eager activity of Sunderland and the bustle and strenuous enjoyment of the brand-new watering place, Roker.

Forty-five miles away from our lofty station is the North Sea, and on a clear day we may catch the gleam of its waters. On our right hand the head waters of the Tees are rushing towards Caldron Snout—"the kettle spout"—at the exact junction of Westmorland, Yorkshire, and Durham. A little footbridge a short way above the rapids enables one to pass from Durham into Westmorland, and after a few steps south one is in Yorkshire. To the left the Derwent is beginning its long journey to feed the Tyne; and behind us the Tyne itself twists northward to meet the Derwent many miles away.

Beyond the moors the coal deposit, the richest in Britain, covers two-thirds of the area of the county. Here it is that the villages and towns cluster closely together. For the coal provides the motive power to the works in the coastal plain, to the iron and steel forges, to the huge furnaces that separate iron from the stone of the hills, to the chemical and glass factories, and to the busy shipbuilding yards. The unceasing din in these—of the riveters fastening together the steel sides of great ships of war or splendid liners, of the rattling chains, hissing steam, and hooting signals—well emphasises the industrial character of our county.

3. Size. Shape. Boundaries.

Between the Tyne and Derwent on the north and the Tees on the south the county of Durham is thrust like a blunt wedge dividing Northumberland and York-shire. The apex of the wedge reaches the eastward slope of Cross Fell but stops short of its summit. From the narrow western border, less than ten miles in width, the county widens to the broad base of the wedge along the North Sea. The width here from South Shields at the mouth of Tyne to Port Clarence at the mouth of Tees is almost exactly thirty miles.

Measuring from the point where Crook Burn enters the Tees — the meeting-place of the three counties Cumberland, Westmorland, and Durham—first to South Shields and then to Port Clarence, we obtain almost equal lengths, each about forty-five miles. If now we draw lines from one to another of these three points, we see how the county stretches a little to the north, a little to the east, and a great deal to the south of these lines. The area is thus much more than that of the triangle we have made. The county actually extends over 1014 square miles, or 649,244 acres. It is therefore almost exactly one-sixth of its southern neighbour, Yorkshire, the largest of the counties, which has an area of 3,872,719 acres, and it is just a little over six and a half times the size of Rutland, the smallest of our counties; so that it is midway between the largest and the smallest. It is, however, below the average area of English counties, though it is actually nineteenth in size.

The boundaries of the county are everywhere natural boundaries. That is, they are clearly marked mountains, rivers, or seas, and are not artificial limits that can be known only by means of boundary stones or posts. The Tees, the watershed of the Pennines, the Derwent and Tyne, and the North Sea effectively cut off the county. Beyond the high moorlands on the west are the counties of Westmorland and Cumberland, across the Derwent and Tyne is Northumberland, and south of the Tees is Yorkshire.

We shall travel along the bordering rivers and along the sea-coast presently : we will now trace the land boundaries, beginning on the western side. The wild moorlands, stretching for ten miles from the point where Crookburn Beck enters the Tees to the summit of Kilhope Moor, are part of the topmost ridge separating the eastern from the western waters. It consists of bleak treeless tracts of heather and peat moss, with here and there vast stretches of furze, forming in spring and summer, when they show their golden flowers, beautiful contrasts to the beds of purple heather. The ridge in places rises into greater heights than any elsewhere in our county—Burnhope Seat, Dead Stones, and Kilhope, at the junction of Durham with its neighbours Northumberland and Cumberland. From the ridge flow the numberless streams—burns and becks are the north country names —that find their way into the three rivers of the county. From Kilhope a lofty spur of the Pennines towards the east, past Burtree Fell and Nookton Edge, separates Weardale from Allendale, and declines towards Beldon

Burn, the name of the infant Derwent. The boundary line follows the Derwent to Milkwell Burn. Here, five miles above the junction with the Tyne, the dividing line between the counties takes a northerly direction till near Currock Hill it strikes the Stanley Burn, which it follows to the Tyne, thus enclosing in Durham a triangle of

Burtree Waterfall: a headwater of the Wear

land of about twelve square miles, shut in between the Derwent and Tyne.

No detached portions now belong to Durham, though in earlier days the prince-bishops exercised their double sway over parts of the country beyond Tyne and Tees. To Bishop Cuthbert and his monks an early King of Northumbria granted the town of Craike, not far from

York, together with the land for three miles round.   A
successor of Cuthbert built a strong castle at Craike and
its ruins are still there, but in 1844 the district became
part of Yorkshire.

Arms of Bishop Bek

To the "patrimony of St Cuthbert" belonged also
the land closely associated with his early life and his place
of death, with Lindisfarne and Holy Island.   To this was

attached the district between the rivers Till and Tweed and the North Sea, a district called Islandshire and Norhamshire. The castle of Norham was built on the Borders in order to defend this most northerly part of the "kingdom within a kingdom." The successor of St Cuthbert bought from a later King of Northumbria the portion of Northumberland lying between the rivers Blyth and Wansbeck. This district, called Bedlington-shire, remained with Durham until 1834. Both it and the northern shires are now included in Northumberland.

We should notice here that Edward the First, desiring to repair a wrong he had done to Bishop Bek, made that famous prelate and statesman King of the Isle of Man. For four years the Bishop of Durham ruled not only his own principality but this island in the Irish Sea also. In the coat-of-arms of Bishop Bek, here shown, the crown surmounted by the mitre denotes his double office of priest and king; the sword of the secular prince is accompanied by the crozier of the spiritual ruler; the "legs of Man" denote the honour conferred on him by Edward; and the cross commemorates his creation by the Pope as Patriarch of Jerusalem, an office which made him head of the clergy of Palestine and guardian of the holy places. Thus the arms denote his fourfold dignity of bishop, prince palatine, patriarch, and king.

4. Surface and General Features.

From the map on the cover of this book we may get a clear idea of the surface features of our county. There are in it three well-defined slopes from the mountain mass in the west towards the North Sea. The south portion of the Derwent and Tyne basin inclines with irregularities here and there to the north-east; the basin of the Wear becomes gradually lower from Cross Fell to the east till it ends in the series of low limestone cliffs broken by little clefts that fringe the coast from South Shields to the hook enclosing Hartlepool Bay; the northern half of the Tees basin has a general south-east direction and, in great contrast to the rugged grandeur of the upper basin, borders the sea with the low-lying monotonous salt measures.

The county is on the average one of the most elevated in England. In the west it does not indeed reach the highest part of the Pennines, the summit of Cross Fell, but the whole of the district here, as we may note from the map, is over 1000 feet above the sea. The coast plain is narrow, and even here in the magnesian limestone region we have well-pronounced elevations. Warden Law, for example, south of Sunderland and only two miles from the sea, the meeting-place in former days of the military muster in Easington Ward, is 632 feet high.

A well-defined scarp, an edge of land rising here and there into low hills, separates the exposed coal measures from the newer limestone. From this edge streams run westward to the Skerne and the Wear; and eastward small streams carve their way through the magnesian limestone rock and form the beautiful wooded denes characteristic of Durham, of which Hesleden Dene, Castle Eden Dene, and Hawthorn Dene are the best known and most visited.

The Pennine Moorlands in many parts rise over 2000 feet above sea level. The greatest altitude of all is attained on Burnhope Seat (2452 feet) on the ridge bordering Cumberland. Other elevations in this wild region are but little lower : there is Kilhope Law[1] at the meeting-place of three counties, whence one looks westward over the moors towards Alston or northward down Allendale. There is Highfield above the lead-mines and the expressively-named Dead Stones. An offshoot to the east, a high heather-covered ridge south of the Wear, also has elevations of over 2000 feet, among them Chapel Fell Top (2294 feet) and Three Pikes (2133 feet). Eastward, Pawlaw Pike south of Stanhope rises to 1599 feet, and Pontop Pike, near Consett, over which Watling Street passed into Ebchester, is 1018 feet high. The 1000 foot contour line passes a little to the east of Pawlaw, and from this line to the sea no elevation reaches 1000 feet. Brandon Hill, three miles south-west of Durham, forming the water-parting between the Deerness and the Wear is, however, not far short of this height.

The county abuts on the sea in a series of low limestone cliffs, not very bold or beautiful but most interesting

---

[1] *Law*, which occurs so frequently in the west of the county, is the old English word for a rounded mound, into which the mountain limestone wears.

Stanhope, from the South

and instructive ; steep grassy slopes break the cliffs at
intervals, and through them the numerous short streams
running from the scarp that cuts off the exposed coal
measures from the magnesian limestone have carved the
narrow valleys, often deep and beautifully wooded, which
are called in Durham the denes.   Durham City on the
flanks of the moors may be taken to divide the county

Brine Mills, near Greatham

into two well-marked divisions.   On the west is high
exposed mountain land cut up by deep valleys.   In the
millstone region there is a covering of purple heather or
golden furze, then on the limestone lower down—on
which heather will not grow—there is a sharp change
from moorland to rough mountain pasture and the be-
ginnings of cultivated land.   In this western division it is

bleak and cold in the winter and spring, and the heavier summer rainfall provides the streams with never failing supplies.   In the other division, eastwards from Durham, there is gently undulating or flat land at low elevations, covered in places with thick deposits of boulder clay, the deposits left by the melting ice-fields that long ago ground their way down the valleys.   Great blocks of granite similar to that sixty miles away on Shap Fell are here found, carried by these glaciers far from their place of origin.

It is only along the upper Tees, where the hard columnar basalt with its sharp-cut edges and the peculiar white granular limestone protrude through the softer layers above, that we have really rugged scenery in our county.   The country north of Caldron Snout, bleak and treeless for miles around, is as wild as any in Britain.

Affording a great contrast to its upper basin, the lower basin of the Tees is the only region of the county which we could call uninteresting.   This, the salt district of South Durham, is a region of recent sandstone, and is the lowest and least striking part of the county.   Nowhere else, not even in the mining region with its artificial hills of shale and the furnaces with their attendant heaps of slag, is there any approach to monotony.   Here the dismal landscape is varied by little else than the still more dismal-looking brine pumps.

5. Watershed. Rivers.

Durham forms part of the eastern slope from the great divide that follows roughly the line of highest elevation along the Pennines from the Cheviots to the Peak. The mountain mass in the west of the county, culminating in Burnhope Seat within Durham and Cross

Wearhead

Fell just beyond the borders, is the birthplace of our three main streams. Here within a mile of Burnhope are Crookburn Beck, which rushes south to meet the Tees from Cross Fell; Burnhope Burn, which flows swiftly eastward to meet Kilhope Burn at Wearhead; various little streams that unite to form the Derwent and the South Tyne; and westward the feeders of the Eden.

Two drops of rain falling a foot apart on the slopes of
Burnhope Seat might be carried one to the North Sea,
the other to the Solway.

From the Pennine moorlands two lofty ridges gradually
sinking towards the east enclose the upper basin of the
Wear.   These heather-covered uplands, in places over
2000 feet high, form the water-partings separating the
burns, becks, and sikes that feed the Wear from those
that curve southward to the Tees or northward to the
Derwent.   The slope from the south to the Wear is
narrower and steeper than that from the north, so that we
find the main feeders of the river on the north bank.
With the exception of the Gaunless, which flows along
the base of the ridge, those from the south are short and
small.

A minor divide runs north and south parallel with
the Pennines, at right angles to the spurs from the major
divide.   This lower watershed, the scarp or edge separating the new limestone from the coal measures, stretches
from Sunderland over Warden Law towards Darlington.
From it to the North Sea flow the numerous small
rivers that have carved out through the easily weathering
limestone the beautiful denes characteristic of Durham.
Down the short westward slope run, in a direction
opposite to the main streams, a few small feeders of the
Wear and of the Skerne.

The copious rains on the moorlands and the reserves
of water in the peat bogs on the mountain slopes afford a
constant supply, and the rivers are seldom much diminished
in volume.   At times, however, the Wear and the Tees

in their upper courses swell with great rapidity : the basin is a narrow one, there are a multitude of little tributaries, and when heavy rains fall the narrow bed of the river cannot contain all the water hastening to the sea. Unwary visitors have been washed over High Force by these sudden floods, and at Durham not unfrequently the lower walks

High Force

along the river are covered[1]. As they near the sea both the Tees and the Wear flow in rather sluggish and winding courses, very different from the dash and hurry of their upland waters.

[1] The names Tees and Team, the latter being the stream which flows into the Tyne at Gateshead, mean *spreading out water*—an allusion no doubt to their tendency to flood.

The lower courses of the three rivers have by means of laborious dredging been made waterways navigable for large ocean-going vessels, which are able to discharge their goods and to load with coal fourteen miles from the sea. Because of the industries along their banks and the trade that is carried on by their help, these three short rivers are among the most important in the world. The beauty of the upper streams is no doubt sacrificed, but not without what, to many, would seem adequate compensation—the increased aid of the rivers in the production of the things needed by man. From Gateshead, Chester-le-Street, and Stockton to the sea the rivers are really great water streets running through a busy and crowded industrial town. Actually to the point where the rivers merge into the ocean their banks are lined with quays, workshops, engineering yards, and stocks on which the skeletons of steel vessels are being clothed.

The Derwent flows for most of its course along the northern border. Its head streams, Beldon Burn and Nookton Burn, rise in the lofty ridge that runs eastward from Kilhope. It passes Edmondbyers Common, where the botanist even yet may be rewarded by the discovery of rare flowers or ferns. Below Muggleswick the limestone is exposed : the river has carved out a deep channel through steep banks and winds through picturesque woodland. At Consett with its huge ironworks loneliness and charm are left behind except for a few favoured intervals. One of these is the passage through Chopwell woods, where it leaves the border, and with the Tyne encloses a peninsula, well-wooded and picturesque to the west but presenting on the east the typical scenery of the mining region. It joins the Tyne at Derwenthaugh[1], where of late coaling staithes have been built. Just before the Tyne touches the county of Durham it passes the little colliery village of Wylam, famous as the birthplace of the father of railways. The house in which George

The Tyne at Howdon

Stephenson was born, a little two-storied cottage, stands alone close to the railway.

Dunston, just below the junction of Derwent and Tyne, has within a few years become an important coal-port, and owing to many new industries—there is a monster flour-mill and a host of small foundries—has a rapidly increasing population. From Gateshead to the

[1] A haugh is the land reclaimed from river or sea.

sea the Tyne is the greatest industrial waterway in the
world. It is, in fact, an almost continuous tidal dock lined
by chemical works, shipbuilding yards, engineering
sheds, and gigantic coal-shoots. The River Tyne Commissioners have deepened the river so that even the
Mauretania found safe passage down it from her birthplace.

From the ravine between Gateshead and Newcastle
and crossed by bridges—the Swing Bridge on the low level,
the High Level, and the new King Edward Bridge—that
are splendid examples of modern engineering skill, the
Tyne flows twelve miles to the sea ; and the whole way
is marked by man's activities. Hebburn with its marine
machinery, its torpedoes, its boiler-making, ship-building
and repairing is typical of the other towns. Then we
come to Jarrow with its old-time memories of Bede, and
of the ruined haven Jarrow Slake. But now the Slake
has been deepened, great timber ponds have been constructed, there are famous ship-yards—the *John Bowes*,
the first screw collier, was built here, and in modern days
the turbine engine was here perfected—and where was
once the quiet home of Bede, the flames of the furnaces
look strange in the night and the din of industry rarely
dies away. Adjoining Jarrow is Tyne Dock, of which
we speak in the section on shipping ; and forming one
town with Tyne Dock is South Shields, its lengthy pier
marking the mouth of the river.

The Wear, the special possession of Durham, for it
touches no other county, is as interesting as any river.
It has lovely stretches of scenery in its middle course,

among the mountains it is wild and romantic, and nearing the sea it flows through scenes of strenuous activity which are rarely equalled. Old history and modern enterprise combine to make it noteworthy. It rises on the western border, where Kilhope Burn passing over Burtree Falls, and Burnhope Burn, fed by moorland streams from the "Dead Stones," unite at Wearhead. Here, many miles from the summit of the mountain ridge, the railway which covers the plain below with a cobweb of lines terminates ; the higher land is tenanted only by wild creatures and by black-faced sheep and their shepherds. Four miles past St John's Chapel the Rookhope Burn, and at Stanhope the Stanhope Burn, enter the stream from the north. These flow through the wild "forest," as waste and solitary as when the Bishop and his train hunted there. Past Frosterley and Wolsingham the Wear reaches Witton-le-Wear, round which is found the most delightful river scenery. Here is the junction of the lead measures with the coal-field. At Bishop Auckland, where the Bishop has his one remaining castle, the Gaunless from the southern ridge joins the Wear.

At Auckland the Wear makes a sudden bend to the north and north-east, past Binchester (where the Roman road crossed) and south of Brancepeth to the city of St Cuthbert. Just before entering it receives from the north the Browney, which has come from Lanchester. From Durham it flows, still beautiful, through Cocken Woods, full of wild daffodils in the spring-time, past the doubly ruined Finchale Abbey. At Chester-le-Street, though we are within easy distance of Lumley and

Lambton Castles, we are in the midst of the northern
'Black Country." Coal-dust and engine-smoke are with
us all the way to Sunderland and the sea.

The Tees among the fells is the most romantic of
rivers ; near its mouth, were it not that its marvellous
industrial and commercial activity supplied interest, it
would be one of the most dismal. It begins in Cross Fell
within the Cumberland border as a clear rapid stream ; its
deep sluggish current ends in the sea between stretches
of mud-flats. As far as Middleton it flows through a
country practically deserted ; nearing the sea it passes
by a population as crowded as any to be found in our
land.

From its junction with Crook Burn, where three
counties meet, to the sea the Tees forms the southern
boundary of Durham. From Crook Burn to Caldron
Snout the county of Westmorland marches with Durham
for a distance of five miles ; thence to Port Clarence
Yorkshire occupies the southern bank.

Just before it rushes through the gorge at Caldron
Snout the Tees fills a deep and long valley in the hills.
This winding lake, deserted and still, the surrounding
mountains protecting it from all the winds that blow, is
called the Weel. The cataract where Yorkshire meets
Westmorland is the loneliest and perhaps the finest in
England. The Tees, hemmed closely in by bare black
cliffs of basalt, dashes down 200 feet within the space of
half a mile in a series of cascades. At High Force, three
miles below, the river falls perpendicularly over a rock
seventy-five feet in height.

The Tees at Barnard Castle

At Middleton the Tees ceases to be bound in by the hills and runs over a pebbly bed, broad and open.   To Piercebridge past Barnard Castle and Gainford is the prettiest part of the river.   At Croft it receives its main feeder from the Durham plain, the Skerne, which on its way has passed through the railway town of Darlington. Yarm, lower down, is partly in Durham, partly in Yorkshire.   At Stockton, with its docks under the control of the North Eastern, business begins.   Port Clarence, whence one looks across the river to the multitude of blast furnaces at Middlesborough, is reached; and through the Seal Sands a channel kept deep only by ceaseless dredging leads past the Snook to the North Sea.

Besides the three great rivers of which we have spoken many of the denes along the coast have their smaller streams.   The best known and prettiest are Castle Eden Burn and Hawthorn Burn.

Few of the stretches of inland water in Durham can be dignified by the name "lake."   There are, as one would expect, great reservoirs where the waters of the moors are preserved; there are pretty artificial pools in many of the parks; and there are some strange depressions, at Blackwell near Darlington, called the Hell Kettles, due to the sinking of the land by reason of brine pumping.   But nowhere is there a true lake.

6. Geology.

How important the study of Geology is to the people of a country is easily seen. A knowledge of the soil is of greatest service to the farmer; and on the character of the rocks beneath depend not alone mines and minerals but actually the build and scenery of the very country itself, and the nature of the vegetation. In a county like Durham, where the staple industry is mining, a knowledge of its geology is of special importance.

A visit to one of the limestone quarries or to the low cliffs that fringe the coast will enable us to understand what the geologist means by strata or layers. We see the working surface of the quarry or the bare face of the cliff showing the rocks arranged, not in a disorderly heap, but in beds one above the other. These beds or layers or strata are not often found in a quite horizontal position, though no doubt at the time of their formation—whether by the cooling of molten rocks or by slow deposits under water—-they had usually a level surface. They are often tilted considerably, or are in smooth curves like those we make when we bend a piece of paper. In some instances the forces twisting and crumpling them have been so great that the layers once horizontal have been tilted into an almost upright position. And sometimes the pressure has been sufficient to fracture the layers so that they are no longer continuous ; and into the cracks molten rocks from great depths have been forced.

How the changes have come about we may understand in some measure by noting how the skin of a drying and therefore shrinking apple becomes covered with wrinkles. So the cooling of the molten earth, its consequent shrinkage, and the work of water and weather on its surface through untold years have brought about the present arrangement of our rocks.

From his study of this arrangement a geologist can give us information of what took place centuries before written history began. He tells us that the county of Durham, like the rest of Great Britain, was many times raised and lowered by movements of the earth's crust. Many thousands of years ago there must have been an ice age, and the glaciers, beneath which there can have been no vegetation, slowly ground their way from the mountains to the sea. Long parallel scratches made by sand and grit held in the ice may still be seen in some of our mountain valleys, and huge boulders, having fallen on the moving ice, were deposited far away from their place of origin when the ice melted.

Long before the age of ice there must have been a warm period when a luxuriant vegetation, chiefly giant ferns and mosses, covered the land, and it is to these growths of the far-off ages that we owe the coal deposits that are found over two-thirds of our county. Long after, a great uplifting of the rocks raised a ridge that has now been worn down to the Pennine Chain.

The gradual crumbling away of the deposits by the action of rain, wind, frost, and ice laid bare the older rocks; so that nowadays these latter are often found at greater heights than the rocks more recently formed. As the land again slowly sank beneath the sea the formation

of newer rocks than the coal strata went on; the debris
of the old rocks gradually accumulated until there were
formed under the water layers of mud, sand, etc. which
by increasing pressure ultimately became solid rock.

There are thus four chief kinds of rock surface in
Durham. On the mountain moors are the Carboniferous

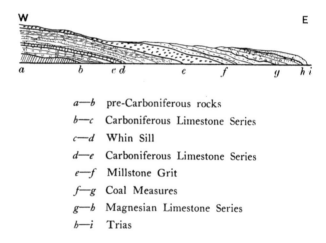

a—b pre-Carboniferous rocks

b—c Carboniferous Limestone Series

c—d Whin Sill

d—e Carboniferous Limestone Series

e—f Millstone Grit

f—g Coal Measures

g—h Magnesian Limestone Series

h—i Trias

Diagram Section across Durham

Limestone and the Millstone Grit; nearer the sea are
the Coal Measures, in places having a thickness of over
2000 feet; fringing and overlying the Coal Measures
is the later Magnesian Limestone; and in the south-east
angle of the county we have the low red-soiled region
of Keuper Marl. We may roughly picture it in this

manner, making a perpendicular cut from Cross Fell
to the coast. The strata all show an inclination to the
east, as we should expect from the position of the Pennine
heights.

It will be noted that in our county the rocks are
comparatively recent. There are only a few small areas,
of which we shall speak later, where the very oldest
rocks have burst through. Even the Mountain Limestone is a stratified rock, laid down in beds under water,
not like granite and other igneous rocks. All our rocks,
with the exceptions noted in the next paragraph, belong to
the division of derived or sedimentary rocks : they are the
off-scourings of the older crystalline rocks. We notice from
the table given overleaf that the first three, the Mountain
Limestone, the Millstone Grit, and the Coal Measures,
belong to the group known as the Primary rocks—that is,
the earliest formed. And all three, being rocks in which
workable seams of coal are found, are called Carboniferous,
or coal-bearing. Again, in the low monotonous stretch
lying between the Tees and a line drawn from Hartlepool
to Port Clarence we have a layer even now in process of
formation, the Alluvium, which belongs to the class third
in order of time—the Tertiary, though this and subjacent
recent rocks are now frequently separated as the Post-
Tertiary or Quaternary group.

The weird romantic scenery of upper Teesdale, with
its rugged black scars—great pillars of basalt not unlike
those of the Giant's Causeway—is due to the breaking
through the Mountain Limestone of the older igneous
rock. The narrow band stretches from Crook Burn

almost to Middleton-in-Teesdale. The river cleaves its way in a deep defile through the basaltic rocks. They occur also in a little band west of Stanhope, where again these hard intractable rocks produce characteristic scenery.

The west is the region of the Lower Carboniferous rocks. The high, bleak, and treeless moorlands, covered indeed in great tracts by purple heather or bright yellow gorse, are the least peopled part of the county. The rocks here, once the lowest, contain the ores of the metals. Lead-mining, now a greatly decayed industry, and iron-mining near Consett, once occupied many men. The Limestone is traversed by many fissures carrying galena, the most important of the lead ores, and zinc blende. Barytes, too—sulphate of baryta so greatly used in glass, porcelain, and colour-work—is found in a fault fissure at Brancepeth. Durham is one of the few counties that produces this mineral in any quantity. The district where the Limestone and the Millstone Grit are exposed extends down the vale of the Derwent to Shotley Bridge; down that of the Wear as far as Witton-le-Wear; and down that of the Tees as far as Piercebridge.

The district of the exposed coal measures covers the lower portion of the Derwent valley, the whole of the Team valley, and the Wear valley from Witton to Hylton. The scenery is here quite distinct from that of the moorlands. Away from the collieries we have a succession of wooded valleys with steep sides, the "denes" of Durham. Between them is good farming land fit for

	Names of Systems	Subdivisions	Characters of Rocks
TERTIARY	Recent Pleistocene	Metal Age Deposits Neolithic ,, Palaeolithic ,, Glacial ,,	Superficial Deposits
	Pliocene	Cromer Series Weybourne Crag Chillesford and Norwich Crags Red and Walton Crags Coralline Crag	Sands chiefly
	Miocene	Absent from Britain	
	Eocene	Fluviomarine Beds of Hampshire Bagshot Beds London Clay Oldhaven Beds, Woolwich and Reading Thanet Sands [Groups	Clays and Sands chiefly
SECONDARY	Cretaceous	Chalk Upper Greensand and Gault Lower Greensand Weald Clay Hastings Sands	Chalk at top Sandstones and Clays below
	Jurassic	Purbeck Beds Portland Beds Kimmeridge Clay Corallian Beds Oxford Clay and Kellaways Rock Cornbrash Forest Marble Great Oolite with Stonesfield Slate Inferior Oolite Lias—Upper, Middle, and Lower	Shales, Sandstones and Oolitic Limestones
	Triassic	Rhaetic Keuper Marls Keuper Sandstone Upper Bunter Sandstone Bunter Pebble Beds Lower Bunter Sandstone	Red Sandstones and Marls, Gypsum and Salt
PRIMARY	Permian	Magnesian Limestone and Sandstone Marl Slate Lower Permian Sandstone	Red Sandstones and Magnesian Limestone
	Carboniferous	Coal Measures Millstone Grit Mountain Limestone Basal Carboniferous Rocks	Sandstones, Shales and Coals at top Sandstones in middle Limestone and Shales below
	Devonian	Upper } Devonian and Old Red Sand- Middle } Lower } stone	Red Sandstones, Shales, Slates and Lime- stones
	Silurian	Ludlow Beds Wenlock Beds Llandovery Beds	Sandstones, Shales and Thin Limestones
	Ordovician	Caradoc Beds Llandeilo Beds Arenig Beds	Shales, Slates, Sandstones and Thin Limestones
	Cambrian	Tremadoc Slates Lingula Flags Menevian Beds Harlech Grits and Llanberis Slates	Slates and Sandstones
	Pre-Cambrian	No definite classification yet made	Sandstones, Slates and Volcanic Rocks

the plough. Here of course is the main coal-mining region, though of late years many mines have been sunk through the newer Magnesian Limestone into the coal measures below.

The Commission which in 1903 examined into our coal supplies estimated that the exposed coal measures extended over 250 square miles, and that underneath the newer rocks 200 square miles could be profitably worked, including the area for ten miles beyond the shore-line. At Ryhope and Seaham Harbour even now the collieries penetrate some miles under the sea. The Coal Measures are almost 2000 feet in thickness, or nearly half a mile. Sandwiched between layers of standstone and shale are about two dozen seams profitable to be worked. In some places the layer of coal is only twenty inches in thickness, in others it exceeds six feet. The average thickness is about four feet. Altogether it was calculated that 5271 millions of tons yet remained below the surface of Durham. At the present rate of output about 130 years will pass before the field is exhausted.

The Magnesian Limestone region is separated from the Coal Measures by a well-defined edge, and terminates on the shore in a series of bold, though not very high cliffs. It is mostly grass-covered rolling land, once purely agricultural and pastoral, but now containing many collieries. The area of the Magnesian Limestone is a triangle eastwards of a line from South Shields to Piercebridge and extending thence as far as the coast.

The red region of Keuper Marl between the lower

Tees and the Magnesian Limestone region is the salt-bearing district, and here, at Greatham, the Cerebos Salt industry has its home. The soft monotonous outlines are due to the easily crumbled strata. The fossilised remains of fish and lizard bones and scales are quite frequently met with in the rocks of this region.

The most recent of the rocks, the Alluvium, formed between Tees mouth and a line drawn from Port Clarence to Hartlepool, is interesting as showing us something of the mode in which even old rocks like the Mountain Limestone were formed. Though now exposed high above sea-level the limestone was gradually formed from the shells of very small marine animals; the Millstone Grit and the sandstones were deposited in the seas or the great estuaries from the dust of older rocks; and the measures in which coal-seams are found tell of many alternate elevations and depressions during an age when luxuriant tropical vegetation covered the land

7. Natural History.

Various facts, which can only be shortly mentioned here, go to show that the British Isles have not existed as such, and separated from the continent, for any great length of geological time. Around our coasts, for instance, are in several places remains of forests now sunk beneath the sea, and only to be seen at extreme low water. Between England and the continent the sea is very shallow, but a little west of Ireland we soon come to very deep soundings. Great Britain and Ireland were thus once part of the continent, and are examples of what geologists call recent continental islands. But we also have no less certain proof that at some period they were almost entirely submerged. The fauna and flora thus being destroyed, the land would have to be restocked with animals and plants from the continent when the land again rose, the influx of course coming from the east and south. As, however, it was not long before separation occurred, not all the continental species could establish themselves. We should thus expect to find that the parts in the neighbourhood of the continent were richer in species and those farthest off poorer, and this proves to be the case both in plants and animals. While Britain has fewer species than France or Belgium, Ireland has still less than Britain.

Most of our existing plants and animals must thus have slowly migrated from Europe during the period in which the shelf at the south-east angle of Britain, now covered with shallow waters, was yet dry land. Some kinds however have been introduced much later by man's agency. Many foreign plants and insects have been brought into Durham by means of the ballast used in vessels that come to Shields or Sunderland or Hartlepool. From the miniature hills formed in these ports the plants growing from the seeds brought overseas have spread abroad. The loose-flowered orchis (*Orchis laxiflora*, a bright red-purple bloom of three petals), for instance, is found on Hartlepool ballast heaps but nowhere else in the island. But, as a rule, these species do not become permanently established. Eggs of insects come, too, as stowaways in the foreign timber or fruit, and some of these hatch, and finding the climate suitable and food plentiful, live and multiply.

The subsequent insulation of Britain, as we have said, occurred before all European kinds had found their way into Britain, for we have no trace of many species that would have found the country a suitable home. Yet we should note that the preservation of great tracts for hunting purposes in our land has given the smaller birds and animals safe breeding-places that are denied them on the mainland ; so that, though we have not so many kinds, our beasts and birds are far more numerous than those of Europe.

Some species, too, were killed out of our land— undesirable kinds like the wolf purposely and perhaps wisely, but many innocent and useful kinds through man's recklessness and ignorance. From the story of our own county we may easily understand how this has happened. We know, for example, that at no very remote period great forests, the home of herds of red deer, must have occupied Wear banks ; for in the dredging operations to deepen the river at Sunderland great antlers—many in a good state of preservation— were often brought up. But no deer except those confined in parks like that at Auckland now range the forests, and no wild boars are any longer to be found, though Brance-peth and Brandon, the "path of the brawn" and the "den of the brawn," remind us of their former existence. The eagle no longer has its eyrie in the county, but

Eaglescliffe suggests that these birds were natives of the county in former days, though now they have been exterminated by gamekeepers and bird collectors. The raven, too, is now a rare bird in this county, nearly extirpated in the interests of pastoral farmers and game preservers.

Certainly our county, which has so great a variety of soil and climate, the rainy mountain moorland, the drier fertile plain, the deeply-carved and sheltered denes, and the bleak exposed hill sides, must have produced an extremely large number of different plants and animals. That is, it had, as we may gather from the traces left, a rich and varied flora and fauna. But in the cause of improved farming wild-flowers have been exterminated as weeds ; only in the fastnesses of the western uplands did the wild cat and the badger find a refuge, and they are now most likely extinct ; the coke-burning and iron-smelting fumes, the smoke and dust from the collieries do not provide an atmosphere favourable to plant growth ; and the pollution of many of the streams has also contributed to kill out, not only the fish, but various species of rare flowers and ferns.

Still, most kinds of British plants are even yet to be found in Durham, perhaps in the sheltered denes or in secluded spots along the river banks or especially in favoured places when we leave the crowded east and travel towards the mountains. Very few of the ordinary flowers and grasses are quite absent. And some species have been noted as occurring only in Durham. Several alpine kinds are found on the left bank of the upper

Tees, chiefly on Widdy Bank Fell; there is *Arenaria uliginosa* (a small five-petalled white flower on a long stalk, known as "bog sandwort") not found elsewhere in Great Britain; *Potentilla fruticosa*, the shrubby cinque-foil, its five yellow petals wide apart on its erect shrubby stem, grows freely, but elsewhere in Britain only in a few spots of Cumberland and Westmorland; there are the tea-leaved willow herb, *Epilobium angustifolium*, and the deep blue spring gentian. Round Edmondbyers the rare little pink flower, *Erinus alpinus*, a curious survival from the Roman occupation, may reward diligent search. It is supposed that this plant, which is found nowhere but in the neighbourhood of the camps, was brought into the county by Spanish legionaries.

In Castle Eden Dene, the largest and most beautiful of all these miniature valleys, there is sheltered a rare and much-prized orchid, *Cypripedium Calceolus*, the "lady's slipper," a pale yellow flower, which grows only on the limestone. There, too, the patient seeker may find another rather local orchid, the narrow-leaved *Cephalan-thera ensifolia*, with its three small white petals, one forming a lip. Other rare flowers and grasses must be merely mentioned: the "bird's-nest" orchis (*Listera Nidus-avis*), found sparingly in Cocken woods near Durham City, the very rare golden saxifrage (*S. hirculus*) on Ireshope, the alpine penny cress (*Thlaspi sylvestre*) in the lead-mines district, the wind grass (*Aira flexuosa*) with graceful panicles on tall red stems, the marsh violet (*Viola palustris*) in boggy land on the mountain slopes, and the bird's-eye primrose (*Primula farinosa*) along the Tees.

The blooms characteristic of the county are, however, the purple heather in vast stretches on the mountain limestone and the golden gorse competing with it for place. In spring and early summer the uplands are one glorious blaze of colour. The cranberry, the knotberry or cloudberry, with its large white flower and raspberry-like fruit, and the bilberry grow freely on the mountain slopes.

Owing to the many spots where moisture and shade are plentiful and to its varied surface, Durham is peculiarly rich in ferns and mosses, though some kinds will soon be extinct in a wild state. The grand *Osmunda regalis*, or royal fern, once plentiful along the Derwent, is now seldom found. At the Falcon Clints, those remarkable cliffs on the upper Tees, *Woodsia ilvensis*, a delicate little fern with very narrow fronds, in England found elsewhere in Westmorland alone, may yet be seen. There, too, in the many chinks and ledges is a surprising assemblage of rare mosses.

The fox, jealously preserved for the famous hunting packs—Braes of Derwent, Marquis of Zetland's, North Durham, and South Durham—is the commonest of the carnivorous wild animals, and the only one in large number. A few badgers may perhaps be left and there are otters in Teesdale. The stoat is common on the western moorlands, and examples both in the dark summer coat and the thick white winter coat are often seen. The common seal was once frequent, and a large colony bred on the Seal Sands at Tees Mouth, but the industries of the coast have long ago driven away these shy creatures.

Rabbits and hares are plentiful in the more secluded parts of the county.

The smaller British birds are numerous in Durham : the game preserves and the many wooded parks afford safe harbourage for their nests. Even the brightly-coloured kingfisher may be seen along the Wear and Tees ; and so can, less frequently, the dipper or water-ousel, though

Eider Duck and Nest

as a supposed devourer of fish-spawn it is relentlessly persecuted and will soon, no doubt, disappear. The nightingale however does not come so far north as Durham. The larger birds, especially the birds of prey, which once found congenial habitation here, have nearly all been killed or driven away. The kestrel hawk may sometimes be seen, but the small merlin common a few years ago is now exceedingly rare. The tawny owl is the only kind of owl at all abundant. There is a heronry in the park at Raby Castle, and this appears to be the only one in the county. It is interesting to note that the eider duck—Saint Cuthbert's bird it is called—on occasions visits the coast in winter from its breeding place in the Farne Islands. The coast gives little protection for sea-birds.

The moors are said to be the best stocked with game in the country. Red grouse, heavier than those of Scotland, are in great number; but the blackcock, once numerous, is now nearly extinct, as is the dotterel. The wheatear, however, the curlew or whaup with its shrill cry, and the golden plover still find homes on the deserted moors, the latter breeding abundantly.

One interesting relic of a species long extinct may be noticed. During the spring of 1878 there were found in one of the sea-worn caves at Whitburn many bones and other remains of primitive man. Five human skulls, bones of the red deer, the roe deer, the badger, and the marten were identified; and, most interesting of all, the skeleton of the great auk, the flesh of which had so long ago furnished a meal for these long-vanished cave dwellers.

The modern conditions of Durham evidently do not encourage butterflies. It is remarkable how many kinds appear to have quite vanished from the county during the last thirty years or so. Out of 35 recorded species there are most likely less than half now to be found; it is most difficult to collect even a dozen kinds. The

Common White butterfly is found everywhere except on the high moors; indeed, it seems to flourish close to the towns and villages and haunts the small gardens near the houses. The Green-veined White too is common in country lanes and woods. The Dark Green Fritillary (*Argynnis aglaia*) can be captured in the Wear Valley; the Silver-washed Fritillary (*Argynnis paphia*) is almost extinct but may, very rarely, be found in Castle Eden Dene. The Small Tortoiseshell ("King William" it is called) is fairly common and so is the Red Admiral, but both the Orange-tip and the Clouded Yellow are rare.

The pollution of the streams by ironworks, blast furnaces, and by the sewage of towns has deprived the lower reaches of the streams of the wealth of fish they once possessed. The trout is rare and the salmon seldom caught. There are some trout, however, in the becks that feed the upper Wear, and above Shotley Bridge the Derwent is well stocked with trout and grayling. The dace (or skelly) is frequently taken from the Tyne; and the greedy pike pursuing the roach or gudgeon in the more secluded parts of the rivers still continues a precarious existence. Eels are quite numerous and appear even to thrive in the polluted streams. Of the deep sea fish we shall speak in the section on Fisheries.

8. Peregrination of the Coast.

The Durham coast is not, like that of Yorkshire, particularly beautiful, nor, like that of Northumberland, wild and romantic. But, throughout its stretch of

45 miles, it is of unusual interest; and, in its many charming miniature valleys—Castle Eden Dene, Hawthorn Dene and others—it has here and there beauty spots hard to equal. At Marsden and at Blackhall great masses of the limestone rock stand out boldly and afford an agreeable variety after the tamer scenery on the rest of the coast. Only one part, however, is low and monotonous, that from Crimdon Beck southwards: desolate sand-hills and rabbit warrens line the coast till towards Port Clarence the dunes are succeeded by dismal salt marshes. At the three great ports—South Shields, Sunderland, and Hartlepool—are the artificial hills of ballast that add to the interest if not to the beauty of the scenery. South Shields is in fact built in great measure on the huge heaps of gravel and sand brought by vessels seeking cargoes of coal, and is as uneven a town as one can find. We should note, however, that these heaps are not likely to grow much larger: many vessels now carry water-ballast, and with the improved organisation of trade a ship comes to port less frequently in ballast.

The general aspect of the coast is that of steep grassy slopes alternating with bare limestone cliffs and broken by the outlets from the great river valleys and the denes. The Magnesian Limestone between Tyne and Wear approaches to typical cliff scenery, rugged and grand. The limestone of which the cliffs are composed is most variable in its nature: in one place it is firm and compact, flaggy and massive; in another it is cellular and brittle, easily disintegrated by water and weather. There is thus a constant difference in the degree of resistance to the wearing forces, whether mechanical—the hammering of the pebbles driven by wind and wave, and the suction of the water—or chemical—the separation of the parts soluble in water. We can therefore grasp the reasons for the strange features of the cliffs. Here and there are old caverns, evidently at one time waterways through

Marsden Rock

the rock; but as the wearing of the base proceeded the roofs or vaults have collapsed. The fallen angular masses have then become cemented together by dissolving lime-stone matter and now appear as solid blocks; while on the other hand passages are drilled by the water through the less resistant part of the rock. Isolated stacks of the more compact limestone stand out boldly : the curious scar called Lot's Wife, near the cave-drilled islet Marsden Rock, is perhaps best known.

South Shields together with the pretty village of Westoe fills the peninsula in the north-east extremity of the county. The old Roman station is now a busy seaport and industrial centre, but yet attracts increasing numbers of holiday-makers. The honour of inventing and of first establishing a life-boat belongs to this place. In the September of 1789 the *Adventure* of Newcastle stranded on the Herd Sands in the midst of tremendous breakers; and the crew, exhausted by cold and fatigue, dropped one by one from her rigging, while thousands of spectators were unable to give aid. The distressing sight led to the formation of a committee to procure a boat capable of living in rough sea; and a boat was ultimately adopted built on the models of two men, Greathead and Wouldhave. This first lifeboat, the *Tyne*, which has been the instrument of saving some hundreds of lives, is still shown on the promenade. On the parade is a fine monument to Wouldhave, " the inventor of the life-boat "; and in the old church of St Hilda the epitaph on the tomb of Greathead describes him also as " the inventor of the lifeboat." No doubt each has a right to share in the honour : Wouldhave's ideas were certainly made use of, though the actual builder of the first life-boat was Greathead.

A little to the south of Shields is Whitburn, a village which is now rapidly becoming a town. Its pleasant position near Souter Point, and the ease and speed of access from the business centres, Sunderland and South

Shields, make it an admirable residential suburb. On Souter Point is a lighthouse warning vessels from the dangerous fringe of cliffs known as the Whitburn Steel, which has a heavy death-roll. Boldon, very different now from what it was in the days when it gave its name to the great register of the bishop's revenues, stands a little inland from Whitburn. Then it was a peaceful farming village, containing twenty-two tenants, who worked three days in every week for their Lord-Bishop, provided two hens and ten eggs, and harvested the lord's crop—in return receiving "a corrody" (the modern "crowdy"). Now the district around is undermined, and mechanics and miners fill the place of farmers.

Whitburn Bay—around which are ranged those parts of Greater Sunderland known as Whitburn, Fulwell, and Roker—is a picturesque curve in the limestone cliffs. Sunderland itself stretches for half a mile north and for over a mile south of the Wear mouth. We speak elsewhere of the industries and trade of this, the commercial capital of Durham. Two striking memorials erected on the sea-front to very different men may be mentioned here. One is a finely carved cross in memory of Bede, the man of thought, who studied and wrote during many years at Monkwearmouth. The other is a spirited sculpture showing Jack Crawford, the man of action, a native of Sunderland, nailing the colours to the mast at the battle of Camperdown. Roker beach and its fine cliff-scenery are easily reached even from the heart of the city, and on summer evenings the beach is an animated sight.

Seaham Harbour, three miles south of Sunderland, is quite a modern town, and is an instance of the establishment of a centre of commercial activity under a carefully thought-out plan.   It did not, as most towns, grow of itself into greatness ; it was founded in the first instance simply as a coal port whence the product of the Marquis of Londonderry's collieries might be exported.   And it

The Cliffs at Roker

was as late as 1828 that the foundation stones of the north pier, and of the first house in Seaham town, were laid.   There is now an admirable harbour on the coast, which between Sunderland and Hartlepool was once a menace to ships ; and 16,000 people now dwell where less than a hundred years ago were bleak uninhabited cliffs.   The town is built on a definite plan, in the form of a crescent in front of the harbour and extends about a mile along the heights of the coast. A long line of dreary crags, broken by a disjointed mass of rock at the entrance of Dalden Dene, stretches northward ; and to the south the Little Stack is a similar break in the monotonous line. In the decade between the census of

North Dock, Seaham Harbour

1901 and that of 1911 the population increased fifty per cent. Seaham Hall in the village of that name was the home of the Milbanke family ; and it was there that the poet Byron married a daughter of the house.

The mouth of Hawthorn Burn leads into Hawthorn Dene, a delightfully wooded retreat still, though the mines are encroaching upon it. The little village of Easington, which gives its name to the ward and the deanery, stands close to the Burn. The church, of Early English construction (about 1270), conspicuous on an elevation above the town, is a sea-mark for mariners. Here it was that Bernard Gilpin, the "Apostle of the North," was once rector.

Horden Hall at the head of a little dene south of Easington is famous for its internal decorations. It is a small building with gables east and west and a projecting porch to the south, and was built probably about 1600. Castle Eden Dene is the largest and most beautiful of all the little ravines that break the monotony of the Durham coast. The castle, of which the certain site cannot now be traced, was after the Conquest in the lordship of Robert the Bruce, an ancestor of King Robert Bruce of Scotland. He granted the lands to the monks of St Cuthbert on condition that they should forthwith build a chapel in the dene. Then the castle towered above the woods, the chapel was almost hid on the edge of its little dene, and the few huts of the serfs were huddled together for protection round the mansion of their feudal lord. Now colliery development and the pressure of population have taken from Castle Eden its peace, and threaten its beauty with extinction. Still, however, there are many charming spots in the dene, Gunner's Pool being perhaps the centre of the most beautiful. Castle Eden Dene would seem to mark the northern limit of Danish influence in Durham : north of it we find the streams bearing the Saxon name "burn,"

south they are Danish " becks." Coundon Beck succeeds
Castle Eden Burn.

Passing Black Halls Rocks and then low sand-dunes
we reach the hooked piece of land that encloses Hartlepool Bay. At the end of the hook the Heugh Lighthouse
stands, the farthest eastward building of our county,
though Seaton Snook is the farthest eastern point.

Black Halls Rocks

Of Hartlepool, the ancient port where Bishop Pudsey
prepared his galleon for the crusade, and of West Hartlepool, the modern town of mushroom growth, the result of
capitalist enterprise, we shall speak at length in the sections
on Shipping and on Industries. But one cannot help
again remarking on the great contrast afforded along this
part of the coast between the old placid days of agriculture

and the new strenuous activity of mining and industrial pursuits.

The traveller by the coast railway between Seaham and Hartlepool will notice that coal is working a transformation here. At Horden is a great modern colliery turning out 3000 tons of coal a day ; Easington Colliery will soon be producing as much ; and at Black Halls the sinking of a shaft is proceeding. Nearing West Hartlepool the vast timber storage grounds are seen, and a fifty acre "timber pond" on which float large baulks in the form of rafts. Smoking chimneys, clouds of exhaust steam, stocks cradling half-built ships, masts, yards, funnels, coaling staithes, and great electric cranes, are crowded together before him. And far off the massive square tower of St Hilda's church, 700 years old, overlooks from a bold headland this animated scene and a great expanse of land and sea besides. In East Hartlepool the ancient gateways and the walls that one Bruce built and another besieged are yet standing, and many of the streets have an old-world appearance. In West Hartlepool is no building earlier than 1847, when the first harbour and dock was planned, and the sand and meadow began to be changed into a bustling seaport.

South of the Hartlepools is the seaside resort of Seaton Carew, unlike many watering-places a village of antiquity. It is built in three sides of a square round a village green, one side of which is open to the sea. Probably the fourth side has been carried away by the eroding action of the sea ; for it is very likely that the site of a Roman road from Seaton to Hartlepool is now submerged. At the point called Seaton Snook, where land is rapidly being reclaimed, we reach the mouth of the Tees.

9. Coastal Gains and Losses. Lighthouse and Harbour Works.

In the chapter on the Geology of our county we mentioned the successive periods of elevation and depression that the land has experienced. Geology teaches us that even the hard Mountain Limestone in the west—some of it half a mile above the sea—was laid down under water; for it is composed of the remains of minute sea creatures that lived long ages ago. In other words, the mountain ridge was once part of a newly-raised beach. We know, too, that much of the area now covered by the North Sea was once dry land over which animals and men crossed from the Continent.

After the great convulsions of the earth ceased, slower movements still went on; and they are not yet ended. In our own county parts of the land surface have sunk beneath the waters of the sea. During the debates in Parliament on the First Reform Bill it was asserted that one of the old constituencies was under the waves of the North Sea, covered by the steady progress of the waters westward—yet members for that constituency still sat in the House of Commons. In Durham the loss—coast erosion and coast depression— has not been so serious; but at Whitburn and at the

Hartlepools submerged forests are observable at low water. In other parts the firm land has gained on the sea. At Cleadon, even yet a pretty country village, on the turnpike road between Sunderland and South Shields, common beach-shells of living species are found at a height of 100 feet above the present level of the water. At Hendon and Fulwell, too, are raised beaches of similar kind and therefore newer than the ice period of which we speak in the section on Geology. The " haughs " at the river bends are all of recent deposit, formed by the loam, sand, and gravel scoured down by the running waters.

The most noteworthy point about our county, however, is the extent to which man's efforts have been exerted to enlarge the area, both of land and water, available for some useful purpose. Land is being everywhere reclaimed for the building of wharves and warehouses, and ship-building yards are now to be seen where a short time ago there were dismal mud flats ; and much of the coast, once neither navigable water nor good dry land, is now either deepened for docks or filled up for workshop sites.

The Tees, at the time of Queen Victoria's accession, reached the sea by four shifting channels, and the small ships that entered had to be guided by portable lights. On the bar at low water the depth hardly exceeded three feet, so that only vessels of small draught could reach Stockton. All this is changed. For thirty years the Tees Conservancy Commissioners have been waging war with nature to improve the waterway. About 24 miles of walls line the river and estuary, and the river now scours out a channel which has a depth of 20 feet on the bar at low water. At the mouth of the river two huge breakwaters enclose a commodious harbour; that at the south, which occupied over twenty years in building, is no less than $2\frac{1}{2}$ miles long, the one from

The Lighthouse, Hartlepool

North Gare is little short of a mile. The building of the breakwater has made possible the reclamation of much of the North Gare sands: on Seaton Snook, for instance, there have quite recently been established large zinc works.

The port of the Hartlepools just north of the Tees

is an instance of rapid development, even more surprising than that of Stockton and the Tees; the county borough of West Hartlepool is indeed an example of mushroom growth such as we usually associate with the New World. To supplement the protection afforded by the headland, Inscar Point, a breakwater stretches to the south-east,

Harbour Entrance, Sunderland

so that at all times in the year a safe approach and adequate shelter is afforded. Wide and deep entrances are provided for the docks—that of Victoria Tidal Dock, in which a splendid fish-quay has just been completed, is 200 feet wide—and lands once waste are now converted into great timber ponds, or are the sites of wharves and warehouses.

At the Hartlepools it is the railway company, at
Sunderland it is the River Wear Commissioners, in the
Tyne it is the Tyne Commissioners and the railway
company together, to whom the improvements are mainly
due.   Sunderland claims to have the easiest and safest
harbour access of all ports on the east coast ; two massive
piers, Roker Pier and the New South Pier, defend the
main harbour-entrance by enclosing an enormous area
of water space ; within, granite-faced concrete piers
protect the river mouth and the entry into the docks ;
a sea-lock, 480 feet long and 90 feet wide, enables smaller
vessels to pass in or out of the docks at nearly all states
of the tide ; and a narrow entrance direct from the sea
to the docks is provided south of the piers for fishing
trawlers.   An ingenious system of lights indicates when
it is dangerous to use this entrance.

The mouth of the Tyne is even more indebted to
man's resource and energy for being made a secure, deep,
and commodious harbour.   The Black Middens, a mass
of rock between Tynemouth and South Shields, and the
Herd Sands near South Shields used to be as great a
danger to vessels as any part of our coast.   Two piers
of solid masonry, that on the north half a mile long and
that on the south little short of a mile, enclose within their
huge arms a harbour secure even when a north-east gale
is raging.   The history of the Tynemouth Pier well
illustrates what the waves can do when driven by a
strong north-easter.   The building began over fifty years
ago ; but a deep enough foundation was not obtained,
and not many years elapsed before the pier had to be

rebuilt. In a north-east gale the white water leaps above the summit of the cliffs, and in one such gale blocks of concrete many tons in weight were twisted from their place—"the waves toyed with them as though they had been pebbles," to use the coastguard's account. The later work of this pier has been carried up in concrete from the sea bottom by means of diving-bell and caisson work, and it has been extended seaward much farther than the original pier.

Three miles up the river is Jarrow Slake, before 1856 a vast unlovely expanse of ooze and slime, covered and uncovered at every tide. In that year its reclaiming began and most of the former marsh is now quays, docks, coaling-staithes, and warehouses. The foundation for the dock walls had to be carried down through 60 to 80 feet of mud.

As we note from the map, deep water occurs quite near the coast except in the south, where we have the low-lying Seaton Sands and the Snook lands. The five-fathom line is nowhere far from land, and the Northumbrian Deep intervenes between Durham and the shallow waters of the Dogger Bank. The access to the coast is almost everywhere without dangers, and a multiplicity of lights, flags, bells, and buoys are provided to make the entrances of the three great ports easy and safe to the mariner.

Despite all these precautions wrecks sometimes occur, and means of saving life are everywhere available. South Shields, we have seen, was the birthplace of the lifeboat. Now we have stations all down the coast, at Shields,

Whitburn, Sunderland, Seaham Harbour, Hartlepool, and
Seaton Carew.   Besides these there is at each coastguard
station a rocket apparatus always ready for use.

10.   Climate and Rainfall.

We need to notice several things when we investigate
the climate or general weather of a region.   We must
discover how great an amount of heat it receives, either
from the direct rays of the sun or by means of winds that
have blown from warmer regions ; and we must note,
too, whether this amount is spread evenly over the year
or whether the heat is much greater at one period than
at another.   We must by our rain-gauge measure the
rain that falls, and notice in addition how the fall is
distributed.   Does most of the rain come during one
season, as it does in India ; or is the fall much the same
in each month ?   And is the fall convenient or not for
the crops ?   Apart from the rainfall, the amount of
moisture in the air is of importance ; and so is the
cloudiness or clearness of the sky.   Two regions may
have the same amount of heat ; but in the one case it
may be due to warm winds and ocean currents, in the
other directly to sunshine.   The second region will be
not only more pleasant to live in but also more favourable
to the growth of crops.

These factors, together making up the weather, depend
on the position of the region ; and three chief questions,
besides many minor ones, must be asked with regard to

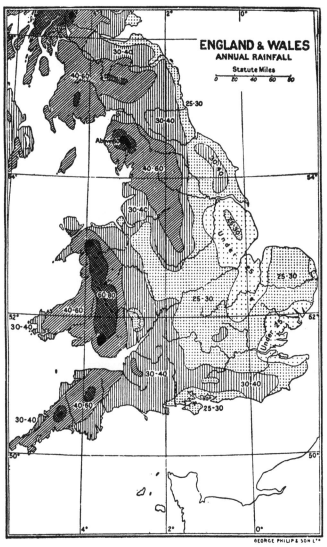

(The figures show the annual rainfall in inches.)

position.   How far is the region from the Equator?
What great stretches of land or of water are near enough
to have an effect on the winds that blow over it?   What
is its height above sea-level?   As a rule, the nearer to
the Equator the warmer the region.   As a rule, too,
regions whose prevalent winds have passed over great
tracts of water have milder weather than do those whose
prevalent winds come from great land masses : they are
warmer in winter and cooler in summer.   Again, the
more elevated a region the colder it is : we may
climb even at the Equator into a region of perpetual
snow.

Let us now consider the conditions affecting our own
country.   It extends roughly from Lat. 50° to Lat. 56° N.,
that is, it is many degrees nearer to the North Pole than
to the Equator.   But its weather is much warmer than
that of other regions of similar latitude.   The prevalent
winds are from the west and south-west, and have blown
over the warmed Atlantic waters.   Our climate is thus
made by them as warm as that of regions much farther
south, and though Durham is less favourably placed than
most of our English counties, it nevertheless participates
not a little in the advantage thus afforded.   Sunderland,
for instance, is in almost the same latitude as Nain,
the chief place, though only a little fishing village, in
Labrador.   But during nine months in the year the ice
shuts Nain from the outside world, whereas Sunderland
is never ice-bound.   In very rigorous winters a thin
covering of ice forms over the docks and appears as a
ragged fringe along the shore ; but we stand in need of

no ice-breaking vessels such as we build on the Tyne for the Russian Government to keep their ports clear.

The average degree of heat during the year in Durham—the mean annual temperature, as it is called—is 47° Fahr., fifteen degrees, that is, above freezing point. But almost as important is this fact: the average temperature of the hottest month, July, is not many degrees above, and the average temperature of the coldest month, January, is not many degrees below, the average for the year. For July it is a little over 59°, for January it is 37½°; that is to say even in the coldest month the temperature is on the whole some degrees above freezing point. We express this by saying that Durham has a small range of temperature, or that it has an equable climate. If we contrast this "insular" climate, as it is called, with the "continental" climate, which we find at such a place as Montreal, the difference is astonishing, for though Montreal is some 600 miles farther south, the mean temperature for January is 25° Fahr., or seven degrees below freezing, though that for July is 70° Fahr.

Yet Durham has a less equable climate than many other parts of Britain. The Western Isles of Scotland are warmer in winter than Durham, though they are cooler in summer; Milford, too, which has about the same temperature in summer, is five degrees warmer in winter, and a snowfall is there a rarity. The main reason for the difference is that the winds from the broad and deep Atlantic have a more moderating influence than those from the narrow and shallow North Sea. And as they pass over the land the west and south-west winds from the Atlantic lose much of their warmth and moisture. The west is, as we see from the rain map, a region of heavier rainfall than the east, the fall decreasing with remarkable regularity, and being greatest during the winter months. Now, when the Atlantic waters are turned into vapour carried by the winds to the land, a great amount of heat is absorbed. When the vapour again condenses and falls as rain this latent heat is released and raises the temperature. So we have a second reason why the west should have an even more equable climate than Durham.

It is the prevailing winds from the west that bring most of our rain. As we see, the rainfall for our county diminishes steadily from the mountains to the sea. In the mountainous part the annual fall averages about 40 inches, in the coast plain about 25. It is therefore on the moors where the Millstone Grit, which does not readily dissolve, is the foundation, that we have the water reservoirs so necessary for the crowded population in the plains. Here for instance are the great expanses, almost lakes, called Waskerley and Blackpool Hill Reservoirs.

The rainfall even in the coast plain is ample for agriculture and pasturage, and the crops grown are limited only by the lack of heat. The months in which most rain usually falls are—as in the case of the greater part of our islands—July and October. On the average the large amount of 3·42 and 3·01 inches of rain falls in Durham during those months. February and April are the driest months. There is not very much difference, the rain being spread pretty uniformly through the year. For the ten years 1881–90 the mean annual rainfall as registered at Durham Observatory, 340 feet above the sea, was 26·11 inches.

Durham is not greatly favoured in its share of bright sunshine. In the wonderfully fine year, 1911, the number of hours registered for the county was 1550, but in most years it is far less. Yet the number of hours the sun is above the horizon is 4450. The moisture in the air forming into clouds and mists often obscures the sun ; and the smoke from collieries, coke-ovens, and blast furnaces often unites with the mist to form a still blacker blanket.

The fogs mostly occur during the winter months. The farmer is fortunate in this respect, that the hours of sunshine come mainly during the spring and summer when the crops are growing and ripening. Yet the fogs are not nearly so frequent as those in South Lancashire and much less dense than those of London, and the keen and bracing east winds often make the atmosphere exceptionally clear. But these east winds are very trying in February and March, and in the high lands to the west the weather during the spring months is bleak and cheerless. Still the mountains in places act as great sheltering walls, so that there is at least one farm-house —in Highfield above the lead-mines—2000 feet above sea-level.

The coast comes within the region of our islands recorded as being least visited by storms. The two chief storm-tracks pass up the west coast of Scotland and up the English Channel. But quite calm days are rare ; and sometimes the east winds are so strong as to make us wonder why the coast should be described " Region of Fewest Storms."

11. People—Race. Dialect. Settlements. Population.

There are surprisingly few traces of its earliest inhabitants in Durham. So few flint weapons and other remains of Prehistoric Man have been found that we must conclude that the district was very sparsely peopled if peopled at all. Its bleak uplands were certainly less inviting than the plains of the south. The little we have to say of the times before the Romans is given in the section on Antiquities.

When the Romans came, the Brigantes, "the people of the summits," the most numerous and powerful of the British tribes, possessed the district we now call Durham. The Roman conquerors included it in their province of Maxima Caesariensis, and since it stood on the effective border of their dominion, took much interest in it. The Angles were the Teutonic conquerors here, and the land between Wear and Tees became Deira, part of the kingdom of Northumbria. Later came the invading Danes, near kinsmen but bitter foes of the Angles, and Durham seems to have specially suffered from them.

The successive waves of invaders have left their traces on the map, and no doubt in the appearance and speech of the people ; but it seems purely guess-work to try to discover these latter. The Celtic words, as we should expect, are still the names of the great physical features : Tyne is *tian*, "running water" and Tees is *tem-ese*, "spread-out water." By far the greater number of names are northern English ; *hope* which enters so largely into the system of names, Burnhope, Kilhope (here the Celtic *kil*, "a church," is united with the Anglian), is a valley running up to the mountain ridge ; *law*, also often found, is the Anglian and Scandinavian for a rounded hill such as the Mountain Limestone wears into. The Danish element in the place-names is not so strong as one would expect ; we often find *beck*, "a stream," especially in Teesdale, where it replaces the Anglian *burn*, and *gill*, "a ravine or chasm," appears in Rowlands Gill, a pretty village south of the Tyne at Blaydon. The Danish termination, *by*, meaning town, is not frequent : it occurs in Aislaby and Killerby. (Butterby, a delightful spot west of Durham city, is not Danish but a curious corruption of the Norman-French *beau-trové*, "a beautiful find.")

The most striking distinction of Durham speech is the peculiar musical rising inflection that marks the end of every sentence. By this rising accent a native of Durham or Northumberland can the most readily be distinguished, and newcomers speedily fall into the manner. The trill on the letter *r*—which is hardly sounded in the South—is another peculiarity. A multitude of words, too, seem to be confined to north-east England. The pitman calls his pigeons *skemmies* ; with the coloured *pase* eggs given at Easter he *jawps*, that is he tries to break and so win his opponent's egg ; he complains to his *marra* or partner of the hard *kyevil*, the portion of the seam in which he works ; and some few years ago, when he obtained his wages each fortnight, he had one pay-Friday and one *bauf*-Friday.

Since the Danes ceased their ravages other foreign elements have been introduced, the Normans who came to hold the high offices in Church and State, the French glass-blowers from Lorraine, the German sword-makers who settled at Shotley Bridge, and even now the commercial interest of foreign countries introduces Germans, Frenchmen, Dutch, even Japanese, into the ports. These elements were small, however, and have exerted little influence on the people.

The population of Durham in 1911, according to the census taken in that year, was 1,369,860, this being an increase of 15·4 per cent. on the population at the census of 1901. The increase is at a lessened rate than that for the previous ten years, though it is greater than for the whole kingdom (9·1 per cent. increase). That is to say, the relative density of the population of Durham is becoming greater. In the whole of England and Wales we have a density of 618 people to the square mile ; in Durham the number is 1350, which is more than twice that for the whole country. Three counties only, Middlesex, Lancashire, and Glamorgan, have more people to the square mile. The actual closeness of the people one to another is, of course, more than these figures indicate. For much of the moorland is almost bare of inhabitants ; on most days we may walk the five miles between High Force and Caldron Snout without meeting a single person or hearing any sound except the startled cries of the curlew, the golden plover, or the grouse. And if we cross over the ridge we reach Westmorland, the least densely peopled county in England, where there are only 82 people to the square mile.

In Durham we find, what is very exceptional in our country, that there are more men than women. For every 1000 men there are no more than 984 women. It is the great mining county and the mining industry has a tendency to draw large numbers of single men into its ranks. During the interval between the censuses 3·1 per hundred of the population had emigrated, quite a large proportion, though by no means so large as that of its neighbour Cumberland. From this latter county over one-tenth (11·1 per cent.) of its population had sought a home elsewhere.

Though the population is so very dense in certain districts it is not massed into very great towns but gathered into very numerous small towns and populous villages, clustered as a rule round the mines. The maps of the colliery districts show this. The largest of the towns, Sunderland, with 151,159 inhabitants, is only twenty-second in population among British towns. But in some districts the villages and towns are clustered so closely together that they are really parts of one great industrial city. From Gateshead, for instance, for twelve miles to the sea at South Shields we have with hardly an interval a succession of crowded towns, Felling, Hebburn, Jarrow, and Tyne Dock.

12. Agriculture — Main Cultivations. Woodland. Stock.

Durham is not greatly concerned with agriculture. For, though fertile land for the plough and good pasture for sheep and cattle are by no means unimportant for us, yet the chief interests of the county are connected with mining, manufacture, and trade. Moreover its farming land is steadily growing less. More and more land is needed for the homes of the increasing population and for their workshops, so that less and less is available for wheat and other crops. Nowadays it is so easy and so cheap to bring wheat into Hartlepool or Sunderland or Tyne Dock from Russia and other countries that the men of Durham find it more profitable to get their wheat by digging coal, or making iron and steel goods, or building ships. Again, as in most of our industrial districts, farmers find a difficulty in obtaining workers.

In the returns of the Board of Agriculture the whole land is divided into three classes. There is the cultivated area, land under crops or permanent grass ; there is rough mountain pasture and moorland ; and there is land on which neither farming nor grazing is carried on, either because it is bleak rock high above the sea, or because it is occupied by works and houses, or because (in Durham to a small extent only) it is woodland.

In our county, as we should expect, little land is covered with corn crops, not one-tenth in 1912. Slightly more is used for the root-crops such as potatoes and turnips, and for clover and other food for horses and cattle. The rough grazing ground, as we noted before, occupies much of the high land to the west. Here a dalesman will tell us that it is not a question of the number of sheep to an acre but the number of acres to a sheep. Actually the very high fraction of one-third of the whole area grows nothing. Contrast this with an agricultural county like Cambridge, where over half the acreage is arable land and much of the remainder grass land, so that less than one-ninth is other than farmed land.

One or two facts must here be noted as modifying the figures to a slight extent. A surprising number of pitmen and mechanics have little garden plots, often in the most unpromising places. These are usually too small to be included in the returns, which take account of such holdings only as are one acre and more in area. But in these numerous small spaces there must be grown a very large amount of vegetables, fruit, and flowers. The hours spent out of the mine or the works are nowadays so many that a considerable amount of leisure is available for the workman, and many therefore take up the pleasant and sensible hobby of gardening. Every little pit village has its annual flower-show, where exhibitors vie in friendly rivalry. In the mining district of Derwent-dale gooseberry shows—or " groser shows " as they are called—are events eagerly looked for, and anxiously prepared for months beforehand.

The diagrams at the end of the book give a much better idea of the relative importance of the crops grown than can be obtained from figures. As we go north in our country the nature of the corn crops changes. Thus, wheat tends to be replaced by the hardier oats. In Cambridgeshire twice as many acres were under wheat as under oats in 1912 ; in Durham the area under oats was about three times as great as that under wheat. The amount planted with barley came between. Beans and peas are but little grown.

The chief root-crop is turnips and swedes ; but the area for potatoes is not much behind. There has indeed been a steady growth in the amount of potato-land, and in 1912 the acres of potatoes—12,889—were 300 more than those of wheat. By far the most important crop is, however, that of the various grasses, clover, sainfoin, and so on. The large number of cows that have to be kept indoors, as one would suppose, consume a great quantity of fodder ; and a great deal is needed for the pit ponies, many of which are born and live below the surface without ever feeding in the open fields.

Less than one-twentieth of the area at the last Ordnance Survey came under the heading of Woods and Plantations. The chief trees are oak, ash, elm near rivers, hazel, alder, and holly. Beech, larch, and sycamore have been extensively planted. Most of the woods are along the river banks, where there is sheltered ground with deep and open soil. Remnants of the native woods that once covered the county almost to the mountains are few ; the chief are in the Derwent Valley. There we ought specially to notice the grand woods at Chopwell. These came into the possession of the Crown when the monasteries were dissolved, and they once provided oak for the ships of war. The first three-decker, *The Sovereign of the Seas*, was built of oak from Chopwell. So were Berwick Bridge and the old Tyne Bridge, though at times the king's builders were obliged to go to Brancepeth " where there was excellent provision of long timber."

As in most parts of Great Britain the increase of

Durham Shorthorns

cattle-breeding, and especially of dairying, seems to have led to a reduction of sheep-farming. One cannot indeed doubt that farmers will more and more find their profit in cattle. There is in Durham an enormous consumption of condensed milk from Switzerland and other countries, of Kiel butter, and of American cheese. Much of this consumption could with great advantage be supplanted by home produce.

The county is known in all parts of the world for its breed of shorthorn cattle. Many of the pedigree animals have realised enormous prices, and have been taken to the Argentine, to Australia, and to New Zealand, in order to improve the breed in those countries. The total number of cattle in 1912 was 76,115, and this is, quite unexpectedly, rather less than the total for 1911. But the decrease is probably due to the unusually hot summer of 1911 and the consequent difficulty of procuring pasture and water, and there is no reason to anticipate another set-back of the steady increase in numbers during the last few years. Of the whole number of cattle about one-third are cows giving milk.

In spite of the displacement of horses by mechanical means of moving loads, the number in Durham does not decrease rapidly : in 1912 there were 23,343, a few less than in the previous year but actually more than in 1910. The horses reared are mainly Clydesdales, but shire horses also are bred.

Not much advantage is taken in Durham of the facilities for obtaining what is called a " small holding " of land—five acres or less. Either because the better land is so expensive, or because high wages can be earned in other industries, the Durham worker does not care to leave what seems a sure thing for the risky business of farming. There were in 1911 only 188 of these small farms, a number far below the average for England.

We should note that the Durham County Council

actively encourages the cultivation of the land and the
rearing of stock. There is at Offerton Hall a Dairy
School from which much is expected ; and it would
certainly appear that the dairy farmer with so great a
market at hand should succeed. The famous cheese of
Wensleydale is made also in farms on the north side of
the Tees. Forestry is taught at Armstrong College,
and in 1911, in order to help and advise small allotment-
holders and kitchen-gardeners, the County Council appointed a travelling instructor in Horticulture.

13. Industries and Manufactures.

Durham, the county of multifarious trades and
manufactures, has nevertheless been the home of other
industries which have vanished. The cultivation and
manufacture of mustard, not so many years ago, was
that for which Durham city had a world-wide fame.
There was a common saying that Durham was famous
for old maids and mustard. A less pungent but much
cheaper article has ousted Durham mustard. A pottery
at Gateshead was once kept in constant employment,
making pots for the mustard, and the pottery work of
Sunderland too was of some magnitude, but the German
potteries and those of Stafford have been able to displace
most of the trade. Then there was glass-making. On
Tyneside settled the glass-blowers driven from France by
the religious persecutions at the end of the seventeenth
century. The first window-glass was made at Newcastle

and was used in the windows of Jarrow church ; but the industry, once most important, is now represented by very few firms. A German colony was brought to Shotley Bridge for the making of sword-blades, and at one time Durham swords had a wide reputation. This industry, too, has vanished, as until lately, had the carpet manufacture of Durham city. Now, however, expensive

Palmer's Shipyard, Jarrow

hand-made carpets are manufactured in Durham, and exported to a considerable extent to the United States. And, though matches were invented and first sold at Stockton, the trade has now deserted Durham to find homes in London and Liverpool. The tanneries of Barnard Castle, once so well known, are likewise no longer existent.

No doubt the main reason for the decay of these small industries is the fact that for the very great ones— iron and steel making, shipbuilding, chemical manufacture, and for all the operations connected with trade—Durham has peculiar advantages. The county ranks with the Black Country and with South Wales for its iron and steel work, and the machinery, rails, and other products of the forges find their way to all parts of the world. In the heart of Africa, for example, a wonderful steel bridge spans the Zambesi River with one huge arch : this was made in sections at Darlington and despatched to be fitted together six thousand miles away. The sheep farmer in New Zealand obtains his iron fencing and the railway builder in Australia his steel rails from the mills round Gateshead and Sunderland.

Shipbuilding and engineering—the making of marine engines and boilers especially—are the great industries. Within three miles of Wear mouth there are more shipyards and engine works than are to be found in any equal space in the kingdom. From one splendidly equipped yard a new vessel glides about once a fortnight into the river. To sail up the Tyne from South Shields to Gateshead is to pass up the water street of a vast industrial town ; with hardly an interval shipyards, marine-engineering shops, chemical works, docks and wharves, line the banks. At Jarrow many of the splendid war vessels that compose our fleet have been built, as well as torpedo-boats and torpedo-destroyers. The lowest reach of the Tees is likewise a busy hive of industry, where steel shipbuilding is making great progress.

Launch of H.M.S. *Mary*

A trip from Hartlepool to Stockton makes one understand how it is that half of the steel vessels built in Britain come from these three small waterways.

The building of the ship is by no means the sole industry connected with it. The Tyne, Wear, and Tees are famous for quick and efficient repairing at a low cost, for the supplies of coal and iron and the skill of the workers make the north-east ports admirably suited for this work. Continental vessels and others from much greater distances regularly come to the Wear or the Tyne for repairs.

Moreover the marine engineers of the North-east have a world-wide reputation. It is quite a common thing for vessels built elsewhere, at Belfast or Birkenhead perhaps, to be fitted with engines and boilers from Sunderland or Shields or Hartlepool. Many of the giant steel castings, too, now needed for the largest vessels, go from Durham. One forge at Darlington makes a speciality of such work, so that we have the curious division of labour by which a town miles from the sea is helping to build a vessel on the coast. How much the world depends on this little county for its shipbuilding material we may realise by noting that there are at Gateshead and Sunderland large works engaged in making nothing but chains and cables, or nothing but ships' anchors, or even the brass-work for engines alone.

The iron and steel works in the county provide not only materials for ships, but bridge-work, steel plates and steel castings, and rails which are sent all over the world, and link together the capitals of Australia or carry the wheat of the Argentine plains to the ports. At Darlington locomotives for the North Eastern Railway are made, and at Gateshead are great railway repairing shops. Another interesting branch of the steel industry is the making of printing presses; many of our chief newspapers are printed and folded by machines built at South Shields. Iron-smelting is on the largest scale in South

Blast Furnaces, Jarrow

Durham, and the furnaces need to be fed with ore imported from Spain as well as with the rich Cleveland ores, and pig-iron therefore, as well as its later stages, is an important item in the list of exports.

The chemical industry, though not so great as once it was, is yet of importance along the Tyne and at Washington. The manufacture of lead, for which purpose the Romans valued the district, no longer stands among the chief industries. Paper-making, especially from wood pulp brought from Sweden, is a rapidly growing manufacture at Hartlepool and near Shotley Bridge, where the pollution of the water has ruined many favourite haunts of the botanist. The abundant deposits of fireclay found under the coal and the fact that it can be so cheaply worked has created a great industry, for export as well as for home use, in firebricks, retorts, and so on.

14. Mines and Minerals.

How very important the nature of the rocks beneath the soil is to Durham we shall readily see from a consideration of a few facts. First and most striking is this, that in our comparatively small county little short of one-sixth of all the coal produced in Great Britain is mined. Not even Yorkshire, so many times larger and with so great a coalfield, produces as much. Durham is *the* coal-mining county. In 1911 the coal-production of Britain amounted to 265 millions of tons, of which nearly 40 millions were mined in Durham.

Coal, however, though by far the most important, is not the only kind of wealth extracted from below the surface. Many of the lead-mines, indeed, which once provided the chief occupation for the dwellers in the west, are now deserted, for so cheaply can lead be brought from abroad that only the most productive mines can be profitably

worked.   Yet two counties only, Derby and Flint,
produce more lead : in 1910 the lead smelted from the
ore amounted to 2840 tons, and the silver found mingled
with it was 17,330 ounces.

Salt, obtained from very deep borings in the red-soil
region north of the lower Tees, is also of great value.
Again two counties only produce more than Durham:
Cheshire, "the salt-cellar of England," is of course first
with nearly two-thirds of the total, then Leicester with
slightly more than Durham, then Durham, where in 1910
there were obtained 175,923 tons of salt, that is to say
about one-eleventh of the whole amount.

Other minerals there are of minor importance.   In
the majority of the collieries the coal-seams have as their
seat-earth or "underclay" the compressed soil that once
existed beneath the heaped-up decayed plant remains of
the marshes, making a firmly-knit clay which is of great
value in the manufacture of firebricks, retorts, pipes, etc.
Some of the glazed bricks and tiles are shipped from
Sunderland as far away as San Francisco ; and there is
a steady trade with the Baltic ports.

As we ascend Weardale we leave the coal region and
enter into the quarrying district.   At Tow Law, where
the inhabitants still point out the remains of the first blast
furnace erected in the north country, at St John's Chapel,
and even high up on the slopes of Burnhope, gannister is
obtained.   This is a very hard, white, compact sandstone
which is unaffected by fire, and is largely used for the
insides of steel-furnaces.

Durham is one of the few English counties from

which barytes—or sulphate of baryta, so greatly used in paint and colour-making—is obtained in any quantity. It comes mainly from a cleavage between the coal-seams at the New Brancepeth Colliery. Then, chiefly in upper Weardale again, the beautiful spar known as fluorspar (the "Blue John" of the miners), utilised largely for the

Coal-miners at work
(*Showing the use of the compressed air coal-cutter*)

making of ornaments, is quarried. Lime is closely associated with the coal, so that we have at hand all that is required for the iron and steel industry. For the iron mines of the Cleveland Hills are near, and into West Hartlepool a principal import is the iron ore from Spain.

In the salt measures is found gypsum—sulphate of lime which when strongly heated becomes the white powder known as " plaster of Paris." The close-grained sandstone of the west provides material for the well-known " Newcastle grindstones," and in 1910 more than seven thousand tons of these stones were shipped abroad from the Tyne. Everywhere there is an abundance of good building stone : the Frosterley quarries, as we have seen, provided the delicately marked stone—" cockle-shell limestone," or " Frosterley marble "—out of which the great pillars in the cathedral are constructed ; and in Upper Teesdale ordinary bricks of burnt clay are quite unnecessary. In Middleton all the houses are entirely of stone, even to the roofs and chimneys.

We will now consider the first three minerals in detail. Coal as the largest employer of labour and the foundation of the other industries concentrated in the coastal plain claims our attention first. The wage-earners (155,113 in 1910) in the coal-mining industry are as numerous as those in the very important metal, machine, and shipping industries all combined. Obviously it attracts other industries into its vicinity ; and, since it forms about nine-tenths in weight of outward cargoes—thus removing the need for departing ships to sail in ballast for our food-stuffs and raw-materials—it helps to cheapen freights on goods coming to England. Moreover, it was the needs of the coal trade that led George Stephenson to invent his locomotive, so that indirectly our wonderful railway system and its offshoots the world over originated from the coal-mining industry. Durham may not have been the birthplace of coal-mining, but the industry is a very ancient one here. In 1180, for instance, we read that a grant of coal-bearing land near Coundon was made to a collier.

We have spoken of the nature and extent of the coal measures in the section on Geology. As the beds dip towards the east, the sinkings must become deeper as the

Cooling Coke at Marley Hill Colliery

sea is neared; and the collieries on the eastern side possess several valuable seams which are absent from those on the west. The coal produced in West Durham is used mainly for coking and manufacturing; East Durham produces in addition excellent house and steam coal. About half of the total output is either exported—Germany being by far the greatest customer, with France and Italy some way after—or sent along the coast to London and the south, or used to feed the furnaces of steamships. The coal-shipping ports are in close touch with the collieries and are admirably served by the North-Eastern Railway, which in some cases owns the docks and coaling-staithes— Tyne Dock, Dunston, and Hartlepool for instance. At Tyne Dock there are wonderful electric shoots which load a ship at the rate of 500 tons an hour, so that we are not surprised to find that over seven million tons are annually sent from this port; and at Dunston Staithes, two miles above Newcastle, where nowadays great sea-going vessels can coal, they are nearly as speedy. From Sunderland in 1911 there were sent almost five million tons, from Hartlepool a little over two millions.

We should note that for the propping of the roofs in the mines, for the laying of lines over which the tubs are to run, and for lining the shafts, an enormous amount of timber is needed. This is the chief reason for the development of the timber import into the Durham ports. West Hartlepool has very rapidly become the fourth timber-port in Britain, and during the summer and autumn vast quantities of wood from the Baltic come into the port.

The price of the coal at the pit head is in general about 10s. a ton. From this the miner receives his wages; there are many other expenses of the mine; and the owner of the surface under which the coal is worked receives about 6d. a ton as a " royalty." In the old days when all the land and its contents belonged to the king, he granted portions away on condition of obtaining a definite share of the produce. What was once paid to royalty is now paid to the landowner.

Electric Cranes discharging pit-props at West Hartlepool

As we should expect, the fields recede as the collieries advance. The burning of coke especially is harmful to vegetation, so that too often the coal industry is associated with gloomy and smoky landscapes; and in the region of iron-smelting the slag-heaps—some still fiery like miniature volcanoes, others sinister-looking with pools of black water—make the scenery even less inviting. The partial removal of the props after a seam has been worked out often enough leads to sinking of the surface in the west, though rarely in the eastern field where the deep workings are. In these depressions water often lodges, so that though lakes are absent ponds are plentiful in the coal district.

Lead-mining is even older than coal-mining. The Romans probably worked lead in the west, and for the fuel to smelt the ore destroyed the forest trees, remains of which are often visible in the black peat of the moorlands. A well-known charter of King Stephen grants the mineral rights in Weardale to his nephew, Bishop Hugh Pudsey. No doubt this grant related to the silver found mingled with the lead, for, as lords-paramount, the Bishops already owned all minerals except the precious metals, which were regarded as a royal perquisite. This is the more probable since at this time a mint had been established at Durham, and the Bishop, assuming yet another kingly function, issued his own coins. Now the lead-mines, once a chief source of the cathedral revenue and the main employment in the upper dales, are, as already stated, but little worked.

The salt industry of South Durham is a most interesting one. The salt-pans of Greatham, where the new

Cerebos works are situated, were famous in the days of Edward the Confessor. The salt deposits are here deep down, not near the surface as in Cheshire. The beds have been produced by the evaporation of ancient lakes or portions of the sea ; the elevation of the ocean floor perhaps cut off a part of the water, and the detached part

Salt works
(*Preparing table salt from the brine*)

would dry up and leave a deposit of salt. Or sometimes an area of the sea became partially land-locked and was flooded at spring tides only. Between the tides evaporation proceeded, and thus in the course of ages thick beds of salt were formed. The salt of South Durham must be of immense antiquity. Many thousands of years must have passed after the completion of the salt bed before a thousand feet of rock—almost entirely alluvial—were deposited over it and, after successive upheavals and subsidences, the salt bed reached its present level.

The salt is won from its deep hiding place by being pumped up as brine. Some of these pumping stations are shown in the illustration on page 18, and very dismal they make the landscape look. The new works at Greatham form, in the flat farming region, an imposing landmark for many miles around.

15. Fisheries and Fishing Stations.

" The greatness of England is built upon the herring and the turnip," said an old writer, meaning that the sea power of our country depended on its fisheries—which provide a splendid school for seamen—and its wealth on the wool of its sheep—for which animal the turnip affords winter food. Its old predominance in the fisheries has not deserted Britain; indeed the number of fishermen, the amount of their catch, and the scope of their operations are steadily, in some cases rapidly, growing. Especially is this so in the Durham coast towns. At West Hartlepool for example, now ranking after Grimsby and Hull as the third of the north-eastern fishing-ports, the take of fish has multiplied five-fold within ten years. We do not nowadays export the wool of our sheep : indeed the factories of Leeds and Bradford are kept busy only by the help of wool imported from Australia and South Africa. And of most foodstuffs, butter and bread and meat, we produce far less than we need at home. Fish, however, we send abroad in large quantities. During the herring season in the autumn of 1911, there went from Yarmouth, the famous bloater town, a hundred cargo steamers, which carried over half-a-million barrels of cured herrings to the ports of the Baltic and the Mediterranean. And to south Europe go also salted pilchards from Cornwall and Devon. The Durham ports, however, have a vast population behind them or easily and speedily reached by its excellent railway service, so that the fish brought into Hartlepool, Sunderland, and Shields finds a ready market at home.

It is a curious fact, and in some ways characteristic of our country, that our Government spends less money on the fisheries than does nearly every one of the fishing countries—Norway, Holland, Denmark, Germany, and France. Yet British fisheries are much more valuable than those of all the rest of Europe together. The east coast preponderates greatly over the west in the fisheries: one reason being that the North Sea is colder than the waters on the west, and the food-fish are as a rule those living in colder water. Another reason is that the shallow parts of the ocean floor, where the fish mostly feed, are more common on the eastern than on the western shores. The Dogger Bank and much of the remainder of the North Sea is one great fishing-ground, but on the Atlantic side the ocean quickly deepens and fish are both less numerous and less easy to catch.

This advantage is not, however, of so much import-ance nowadays; for half of the fish landed in Britain are caught in waters beyond the North Sea. Only in 1905 a rich ground for plaice was discovered in the White Sea ; and from this area in the far north to Morocco in the south fishing is carried on from Hartlepool, Sunderland, and Shields. The methods of taking fish have of late

Fish Quay, Hartlepool

become so much more effective that it seemed likely the supply might begin to fail. So regulations as to nets, distance from shore, close season, etc.—all with a view to protect the eggs and young of the fish so that more would arrive at maturity—have been made. The enforcement of these rules is entrusted in Durham to the

North-Eastern Sea Fisheries Committee, two of the members of which are appointed by the County Council.

Over three-quarters of the fish landed in Durham are caught by trawls, great nets which are dragged along the feeding grounds in the shallow parts of the sea. The species caught in this way are chiefly cod, haddock, lemon soles, and flat fish, and giant halibuts over 400 lbs. in weight have been brought in. Most of the remainder are caught in drift-nets, which hang from the surface in fairly deep water across the path of the shoals. The herring and the mackerel are captured chiefly in this way. Line fishing is nowadays insignificant and accounts for only about one-fortieth of our total catch, though much line cod is brought into West Hartlepool.

The herring, common as it is with us, is found only in the seas of Northern Europe : it does not extend southward beyond the English Channel nor westward beyond Greenland. The shoals appear off Wick in May and move southward followed by the nomadic bands of fishermen.

The coast towns of Durham are admirably situated with regard to the North Sea fisheries and those of districts farther north ; and under the fostering care of the port authorities and the railway company the number of men engaged in this important industry and the amount of fish landed are certain to increase rapidly. The use of steam trawlers has taken away from Yarmouth, Grimsby, and Hull any great advantage derived from being nearer to the Dogger Bank ; and of course the Durham ports are nearer to the northern fisheries—the Lofoten Islands, the

Faroes, and Iceland. At Hartlepool, Sunderland, and South Shields the trawlers are now able to land their catches with ease and expedition at all states of the tide, merchants are able to despatch the fish by express train to the consuming centres, and the vessels may take in stores, coal, or ice without moving from the dock. Excellent facilities, too, are provided for smoking or otherwise curing the fish.

A great deal of sea salmon is brought into Hartlepool and most of it is immediately sent in special fish waggons to London. Cockle-gathering has been for ages an occupation followed by a number of men and women over the marshes at Tees mouth. But much of the fishing-ground nowadays serves more useful purposes than that of being a nursing ground for cockles, and this small industry is being killed by the growth of larger ones.

We must not imagine that the Durham fishing ports, surprising as has been their development, have as yet anything like the amount of business of Grimsby and Hull. Into Grimsby, our greatest fishing port, thirty times as much fish is brought as into Hartlepool, into Hull ten times as much.

16. Shipping and Trade.

On the three north-eastern rivers half of the steel ships of Britain are built; and much of the product of the yards is for the use of the Tyne, Wear, and Tees ports. The astonishing growth of industries on the coal field is more than matched by the development of overseas trade. Nowhere in the world is there such a group of enterprising ports as line the Durham coast from Dunston fourteen miles up the Tyne past Gateshead, Hebburn, Jarrow, Tyne Dock, South Shields, Sunderland, Seaham, the Hartlepools, Seaton, Port Clarence, to Stockton ten miles up the Tees.

By the magnificent breakwaters which have been described in a previous chapter the mouth of the Tyne has been made a harbour where vessels can ride safely in any weather; Sunderland has become one of the easiest and safest ports of access in Britain ; there are no dangers to navigation, and ships arrive in the heart of the commercial centre a few minutes after passing Roker Pier lighthouse. Hartlepool, by lavish provision of docks and of quays with wonderful facilities for loading and unloading, has become the first port of Britain for mining timber; and it is now aspiring to become the great fish port in the north-east. Stockton, too, long content with its ancient history, woke to life within living memory and has made itself famous as a deep sea port and a place of eager business.

Before going into details we should note another general feature. The north-eastern ports are the home of the " tramp " steamer, the earnings of which when plying between places abroad form a great part of our " invisible " exports, paid for by sending to Britain food and materials. A vessel will take a cargo of coal from Dunston to South America, will load with grain in the

Sunderland from the Bridge

Argentine for Rotterdam, will thence take general produce
to the Black Sea, and after nine months will reach Hartlepool with a consignment of wheat from Galatz. This
common carrier for man on the ocean is the special
product of the Tyne, Wear, and Tees. The shipping
trade of Durham differs therefore from that of London
and Liverpool, whence cargo and passenger steamers have
regular runnings on settled routes. It is true that the
north-eastern ports have regular services, to Montreal and
New York weekly, to the Baltic ports weekly, to Smyrna
and the Levant monthly, and numerous others; but the
main feature of their trade is the steamer that has no
regular route and carries on the world's business wherever
the greatest demand occurs.

Sunderland, the commercial capital of the county, is
the port with the greatest trade. In certain branches,
however, it is surpassed by others: Tyne Dock ships
more coal for instance, and the Hartlepools import more
wood and fish.

There is, indeed, the keenest rivalry between the ports,
more especially for the coal-exporting and for the fish
trade. This rivalry is owing not alone to the vigour
and enterprise of the traders, but to the varied ownership
of the docks. The North-Eastern, the largest dock-
owning railway company in the world, controls those at
Tyne Dock and the Hartlepools, the River Wear Commissioners own three large docks at Sunderland, there are
also privately owned docks, and others under the control
of the Town Councils. Remarkable appliances for working cargoes are available—there are for instance electrically-

worked belt conveyers that can shoot into the ship's hold 700 tons of coal an hour, and great steel grabs that take a waggon full of coke and tip its contents, manipulating it as easily as a man might a shovel. Special fish quays are provided, there are facilities for supplying ice and for curing the fish, and rapid carriage by rail is assured. South Shields, Sunderland and Hartlepool are all striving to

Stern frame of the *Britannic* being loaded at West Hartlepool

obtain the greatest share in this rapidly increasing fish trade.

The bulk of the export trade, as it has been for some centuries, lies in the coal and coke. Much is conveyed by colliers to the Thames. It was for this trade that the first screw collier was built at Jarrow; and now at the Pallion Yard at Sunderland there are building self-discharging colliers which would appear in the matter of handling cargo to be as near perfection as can be attained. By a series of belts the vessel discharges its coal into barges alongside at the cost of a farthing per ton, and large numbers of the shore gang are dispensed with. Iron and steel goods come next to the product of the mines. Marine machinery is made not alone for our own ships but for many built in Holland and Germany; and Belfast and the Clyde often procure parts of their liners from Durham works. From West Hartlepool were sent, for instance, the mighty stern frame and rudder-post for the *Britannic*, building at Belfast, and greater still was the frame of the *Aquitania*, now building on the Clyde. A great amount of pig-iron and of steel goods in various stages of working is sent abroad for further manufacture.

Other exports, though much less than the two chief classes, are yet important. Chemicals, mostly manure and sulphate of soda, fireclay goods, lead, zinc, flint glass, salt, and cement, usually form the chief items, but to these allusion has already been made.

The imports are very much less in weight than the exports. The reason for this is obvious. The coal which forms the bulk of the exports is not costly for its weight; the general merchandise, the food and materials sent in return cost much more per ton. Many vessels come therefore mainly in ballast for outward coal-cargoes: at the three chief ports, South Shields, Sunderland, and Hartlepool, there are miniature hills of sand and gravel discharged from vessels that have come for coal but could not obtain freight enough to bring to the county. Some of the small coasting steamers bring back white chalk for the rapidly growing cement manufacture in return for the coal. But nearly always the regular coasters come to the coal ports in ballast ; and so do many of the tramp steamers, which may have landed cargoes at other ports, British or Continental, and are sent to Sunderland or Hartlepool or the Tyne for an outward freight of coal. At other ports there is often difficulty in obtaining cargo : a vessel quickly finds work to do at the Durham ports.

But the vast population behind the ports needs a great amount of food and the industries need a great amount of materials, so that the imports are by no means small. Timber is first. At Jarrow, Sunderland, and Hartlepool there are many acres of water set apart as timber ponds ; and during summer and autumn, when the Baltic ports are accessible, there is an almost continuous discharge of props, deals, and sleepers into the ponds. Most is timber for mines and railways ; but mahogany and other furniture woods are also imported. Iron ore to feed the furnaces is an increasing import : it can be brought so cheaply from Bilbao or Malmö that little is mined in Durham nowadays. The Consett Iron Company, which once worked its own iron mines, now uses imported ore only. Grain of all kinds, more especially wheat from the Black Sea ports, is, as one would anticipate, high up on the list of imports, and at Sunderland and Dunston are gigantic mills, fitted with the most modern machinery. Wood pulp for the paper factories at Sunderland and West Hartlepool, hemp from Germany for the rope-works, Norwegian ice for the use of the fishing trawlers, china-clay from Cornwall, provisions of all kinds—butter from Kiel, bacon from Danish ports, cheese from Rotterdam, fresh and frozen meat, tinned milk from Switzerland, sugar from Hamburg, and eggs in thousands from Havre and Denmark, and a vast amount of fruit—all these swell the amount of imports.

17. History.

We must depend largely on guesswork for the history of our county before the time when it became part of the widespread empire of the Romans. That great people, however, brought it into civilisation, made roads, built bridges, and organised the people, till then bound only loosely together by the tie of common race. The years after the withdrawal of the Romans, about 420 A.D., were full of exciting incidents; and, since the Normans overthrew the Saxon kingdom, the history of the county has been of the very greatest interest. During many centuries it is the story of a kingdom within the kingdom. We have in our county an example unique with us in Britain, though common enough on the Continent —an ecclesiastical state in which the king had little but nominal power. Not "the king's peace" but "the peace of Saint Cuthbert" was maintained by the bishop. The men of the county held their land from the bishop and did him homage; they were "holy-work-folk" (*halywerkfolc*) thereafter, pledged to defend the shrine of Cuthbert but not called on to fight outside the bounds of the bishopric. In other parts of England the right of obtaining minerals was granted by the king, and to him went the "royalties"; in Durham it was the bishop who granted permission and who received a portion of what was obtained. So marked was the separation from the rest of the kingdom that, till the rule of Cromwell, Durham alone of the counties sent no member to Parliament. The Palatinate had a parliament of its own, and the bishop protested vigorously when it was suggested that men from Durham should be summoned to Westminster. "The King's writ is of no effect in the County Palatine of Durham" was the haughty response made by the bishop whenever an attempt to infringe his privileges was made by the monarch.

The bishop had his own mint and his own law courts, and the whole county was studded with his manors. To these, like the king, he shifted his court as business called him. Stanhope Common was his hunting ground and in Boldon Book, mentioned below, we find frequent allusions to the services the bishop's tenants had to render at the hunt. "Plausworth, which Simon holds, renders 20s., and carts wine with eight oxen, and goes in the great chase with two greyhounds"; the "villans of Stanhope make at the great hunts a kitchen and larder, and a kennel; and they carry all the Bishop's corrody from Walsingham to the lodges" (to this day among the miners "crowdy" is a common term for food); other tenants provide ropes or timber or "tuns of beer," and "Ralph the Crafty" trains hawks in return for the holding of Frosterley.

The power of the bishop was at times so great that the king grew jealous of it. In Edward I's reign, for example, the claimants of the Scottish crown, John Balliol and Robert Bruce, were outlawed after their fighting against Edward. Their lands in Durham, Hartlepool, and Barnard Castle were therefore forfeit to the Prince Palatine, to the bishop. But the king, taking advantage of the bishop's absence in Rome, seized the lands and rewarded his own nobles with them. And later, Henry VIII, anxious to diminish the power of his mighty vassal, granted to Newcastle a charter by which the town of Gateshead was included in the important town north of the river. When Queen Mary reigned the town was, however, once more transferred to the bishop. In 1836 Parliament appointed a number of commissioners to look after the business affairs of the church and so to relieve the bishops of trouble about matters apart from the rule of the church. Since that date the bishop has reigned as a prince of the church but not as a great temporal ruler. He has the honour of supporting the monarch at his right hand during the coronation ceremony however, and still retains the coronet in his coat-of-arms.

The Angles at their invasion no doubt found the scattered British in Durham an easy prey. The disciplined and experienced troops of Rome had gone, and the strongest of the British were most likely fighting as legionaries in some distant part of the now diminishing Roman empire, but those that remained no doubt made some resistance against the onward march of the invaders. From the state of the pavements and the number of partially melted coins found, it is supposed that the great

camp at Lanchester was burnt during the struggle.
Ebchester was even more completely destroyed, and the
district shortly became a part of Deira, a kingdom which
by union with Bernicia became Northumbria.

Then, when times were quieter, came missionaries to
these heathen and won them to the faith of Christ.
Some came from Rome, but most came from the Celtic
church which, surviving the onslaughts of the invaders,
had established itself in Ireland and in Scotland. An
offshoot from Iona was Lindisfarne, from which came
Aidan and after him Cuthbert. It is round the incor-
ruptible body of Cuthbert that the early history of
Durham hangs. The great preacher died in 688, giving
his monks charge to carry his bones along with them if
necessity should compel them to leave Lindisfarne, and
during the incursions of the Danes the body was carried
about in obedience to this command, until a permanent resting-place at Dunholm was at length found.
Here round the shrine of St Cuthbert grew up the noble
cathedral. Cuthbert's command: "Tell him to give to
me and those who minister in my church the whole of
the land between the Wear and the Tine for a perpetual
possession. Command him, moreover, to make my church
a sure refuge for fugitives, that every one, for whatever
reason he may flee to my body, may enjoy inviolable
protection for thirty-seven days," had been obeyed by
Guthred, and King Alfred confirmed the grant. The
grotesque sanctuary knocker, now on the north door, the
sight of which must have been welcomed by many a
panting fugitive in bygone days, is well known. And in

the British Museum we may yet see the *Liber Vitae*, the Book of the Benefactors of the Church of Durham. Swiftly the shrine of St Cuthbert, first at Chester-le-Street under the hastily-erected wooden church, then

The Sanctuary Knocker, Durham Cathedral

under the magnificent structure of William of Carileph, became as rich as any in Christendom, and all the land between Tyne and Tees came under the almost unrestricted rule of the head of the church at Durham.

The neighbourhood of Scotland, then and long after an active and vigilant enemy, and the turbulent state of the north, always restless but especially under Norman rule, seemed to require that a power capable of acting with vigour and promptitude when occasion called should exist at Durham. Hence the origin of the temporal power, which custom confirmed. So important did the see become that the kings of England and of Scotland were keenly interested in the choice of an occupant, and conflicts more than once took place before the question was finally settled.

No doubt the power of the bishops grew slowly, and Bishop Walcher, elected after William the Conqueror had visited Durham in 1072, was the first whom we can with certainty call Prince Palatine. William, relying on his support, created him Earl of Northumberland. But Bishop Walcher found his new subjects a fierce and hasty race, and ultimately met his death at their hands. For this the Conqueror's brother Odo was sent with a great force to Durham, and laid waste the county, executing without mercy and without discrimination such of the people as fell into his hands.

It was William of Carileph, who reigned in Durham during William II's time, who in 1093 laid the foundation and built much of the cathedral. The building was continued by his successors, especially Bishops Flambard and Pudsey, till in 1480 the great central tower was complete; and Durham cathedral, "that glorious work of fine intelligence," crowned the rock above the Wear.

Durham Cathedral from the West

One of the most notable among the rulers of
Durham was Bishop Hugh Pudsey, nephew of King
Stephen, who held office for forty-two years. Under
Henry II, who was anxious to have one authority in
the kingdom and heartily disliked the idea of a subordi-
nate independent power, there was constant friction.
Bishop Pudsey would not pay scutage ("shield-money")
to the king at London; he supported the king's rebellious
sons against their father; and is said to have allowed
foreign troops to land at Hartlepool. From the spend-
thrift Richard the Lion Heart, the bishop purchased the
wapentake of Sadberge, and the Prince Bishops were
thenceforth entitled to sit in the House of Lords as
Barons of Sadberge. Pudsey was an insatiable builder.
Elvet Bridge at Durham, Darlington Church, Finchale
Priory, and Sherburn Hospital are a few of the many
structures due to him. But chief of all he urged forward
the building of the cathedral, and he it was who built
completely, and adorned, the beautiful Galilee Chapel at
the west end of the main body, where it appears almost
to overhang the Wear far below.

He was also the compiler of the "Boldon Buke" in
1183. The county of Durham, as being under the
control of its own prince bishop, had not been included
in Domesday Book, the great register of the land compiled by command of William the Conqueror. The
Boldon Book, in which Bishop Hugh Pudsey commanded
to be written his tenants, their holdings and their dues,
is the Domesday of the Palatinate. Boldon, a little village
near Sunderland, was a typical manor of the bishop; the

compilers therefore, having fully described the services and
payments of the tenants there, were usually content to say
of the other manors that the services were the same as
those of Boldon. Hence the name of the register, the
"Boldon Buke." In it we have a vivid and most valuable picture of the life of the people in the Middle
Ages.

All the people were the servants of the lord; and
in return for the work they rendered him, or the provisions they furnished, or later the sums of money they
paid, each had his little holding of land. This provided
for the daily wants of his family. The houses of the
"villans," each in its separate toft, were by the side of
a stream or where the best land lay. Attached to the
village was the common field, where each tenant had
his portion of arable land, divided from his neighbour's
portion by narrow strips of grass. Beyond was the pasture,
where the cattle fed in common under the charge of the
village herd. Beyond that again was the lord's waste
or forest, where the tenants had rights of pasturage, of
swine feeding, and of cutting turf and fire-wood. Each
village had its herd for looking after the stock, its pounder
for taking care of stray cattle, and its smith and carpenter.
Each village provided for its own needs; there was little
exchanging with other districts; and hardly any need of
money.

The modern "copyholder," who holds his land on
payment of the customary services or of the money for
which these services have been exchanged, is the successor
of the "villan," who was for all purposes a serf. His title

to the land is based on a copy of the court roll—of Boldon Buke. He pays to the lord of the manor a "quit rent"— a very small payment—after which he is quit of other services.

Boldon Book shows us, however, that men were beginning to have special occupations. There is mention of a certain collier who "held one toft and four acres for finding coal to make the iron work of the ploughs of Coundon." "Lambert the marble cutter has 30 acres for his services so long as he shall be in the Bishop's service" (no doubt he provided the columns of Frosterley marble with which Bishop Pudsey adorned the Galilee Chapel of the cathedral), "and when he shall have left the Bishop's service, he renders two besants, or 4s." In Wolsingham there were three wood-turners "and they rendered 3100 trenchers" for the bishop's guests; the pounder of Darlington "renders 100 hens and 500 eggs" for his nine acres; "Geoffrey Joic holds 20 acres for 40d.; and he goes on the Bishop's errands." In North Auckland "Gatull the smith holds 16 acres for one pound of pepper and his heirs for 2s. or two pigs."

In chapter 3 we mention the powerful Bishop Bek or Beck and his dealings with Edward I. When the great king, the "hammer of the Scots" died, and Bishop Bek also was dead, the Scots kept the northern county in continual alarm. Once before Bannockburn, and many times after that crushing defeat of the English, they invaded Durham under their king Robert Bruce. One band penetrated as far as Hartlepool and burnt the town, a forfeited manor of Bruce which his own father had fortified. Disease and famine were added to terror of the Scots, so that till Edward III with the help of Durham men won the battle of Halidon Hill in 1333 the county was in a deplorable state.

The Scots did not, indeed, often come into Durham

Neville's Cross

as an army. They were usually small marauding bands. In 1346, however, King David Bruce and his whole army invaded the county only to be defeated at Neville's Cross. From its meeting place in Auckland Park the English army advanced to the defence of the cathedral and Cuthbert's shrine. The Durham men were commanded by the powerful baron Ralph Neville, the lord of Raby and of Brancepeth. Within sight of the cathedral the armies met; the English archers, admirably posted on the hill where the cross of Neville now stands, did much execution; the heavily-armed knights completed the defeat of the Scots, who had not calculated on meeting so great resistance. By evening of that autumn day, October 17th, 1346, the battle of Neville's Cross, as we now call it, was over, and nearly ten thousand Scottish soldiers—nobles and commons—lay dead. Their king was wounded and a prisoner; the Black Rood of Scotland was in English hands, and was presented at the shrine of Cuthbert. A magnificent cross, of which we now see only the remnants, was erected by Lord Neville to commemorate the victory.

Flodden Field in 1513 was the last great fight in this ruinous Border warfare, and there too the Durham men fought in the front and, probably for the last time in battle, the banner of St Cuthbert was displayed. The days of disturbance were nevertheless not quite over. The changes in the religious services during Henry VIII's reign caused in Yorkshire and Durham a rising called the "Pilgrimage of Grace" in 1536. This came to nothing, and though the old system was restored for a time in Mary's reign, the men of the north were still dissatisfied and a more formidable rising took place in 1569. The Nevilles and nearly all the great nobles of Durham assembled at Brancepeth; mass was again said in the cathedral; the one noble faithful to Elizabeth, Sir George Bowes, was besieged in Barnard Castle and after eleven days had to surrender. The idea of the rebels was to release the imprisoned Mary Queen of Scots, to make her queen, and to restore the old religion. But the leaders had spent their money as an old writer says " by doing deeds of hospitality," they were unable to obtain supplies, and the " Rising in the North " speedily collapsed. Little blood was shed in actual fighting but much to punish the misguided people. In Durham city alone sixty-six men were executed.

Durham men as a body took the losing side again in the civil war between Charles I and his Parliament. For a long time the Scots, who supported the Parliament, were stationed at Durham; and during a great part of the war Sunderland was a Royalist port of consequence in direct competition with "the rebellious city of London." Yet the Lord Protector Cromwell in the day of his power planned the establishment at Durham Castle of a northern university " which might much conduce to the promoting of learning and piety in these poor, rude, and ignorant parts." The old universities both protested vigorously against the project, especially against the power to be given of granting degrees. But Cromwell's death, not the protest, prevented the plan from being executed ; and it was not until the lapse of nearly two hundred years that Durham University came into being. The last of the prince bishops gave up Durham Castle to be the home of the new seat of learning, endowed it with a great part of his income, and in 1833 the university first admitted students.

The later peaceful progress of the county, its industrial

triumphs, its wonderful engineering works, its growth in trade, and its attainment of the first place in the world for shipbuilding are told at full length in other sections.

18.　Antiquities.

No remains at all of Palaeolithic men—of men who used rude flint implements—have been found in Durham, and very few of the later Neolithic men who had learnt how to polish and cleverly fashion such stone implements and weapons.　Perforated stone hammers however have been found at Gainford, scrapers of flint at Hamsterley, and polished stone axes at Jarrow.　The county is also greatly lacking in fortified sites and places of burial that can be described with certainty as prehistoric.　The mounds of earth—barrows as they are called—and the stones or cairns that marked the burying-places of prehistoric men are practically absent from Durham.

To the Stone Age succeeded the Age of Bronze, when men had learnt how to work the softer metals, tin and copper.　Of this period remains are more numerous; and one in particular, at the Heather Burn caves near Stanhope, is perhaps the most remarkable find of the Bronze Age ever made in Britain.　Here, in 1812, in a great fissure in the Mountain Limestone, were discovered the home and the whole equipment of a family of this period.　Three human skulls, besides bones of the dog, and of animals that had served as food, were found mingled with a curious collection of bronze implements and ornaments. Among them were tongs and a roughly-shaped copper-waste runner made of bronze. There were swords, axes, and knives with sockets for the handles, rings of various kinds, armlets (one oval armlet was of gold), and bronze discs. There were also carved bone implements, horse and dog teeth pierced for wearing, and knives or razors either for cutting hides or for shaving. A great bronze caldron, 18 inches high and 14½ inches

Oak Canoe found in the Tyne

wide at the mouth, on which there were still traces of carbon, had served as the vessel in which the flesh of the captured animals had been cooked.

The prehistoric canoe in our illustration may belong to the Bronze period. It is made of oak and was found in 1912 buried in sand and silt not many miles farther up the Tyne than where the huge *Mauretania* began its career. The contrast may serve as an indication of the progress in knowledge and skill made by man ; it is an

8—2

advance something like that marked by the difference
between our glorious cathedral and the cave dwelling
just described.

When men had learnt to smelt the harder metal iron
from its ore, we are at the beginning of recorded history.
Traces of men of the Iron Age are scarce in our county,
possibly because iron rusts away so much more quickly
than bronze. An iron sword handsomely shaped and
ornamented was, however, found preserved in its bronze
scabbard at Barmston near Sadberge, and is now in the
British Museum.

Roman relics, as one would expect from the importance of our county in the Roman period, are plentiful,
not alone of their great works—roads, bridges, and fortified
camps—but of their altars, coins, pottery, and weapons.
Many of the altars are in the wonderfully interesting
museum in the Cathedral Library; and in the public
library at South Shields there is a remarkable memorial
stone, erected by a Roman officer in honour of his British
wife. Little is now left of the fortified camps. That at
Chester-le-Street is covered by more modern buildings.
The western ramparts of Binchester have been borne
down by the river Wear. The greatest of all, the camp
on the hill at Lanchester, may indeed be traced throughout
its defensive walls, many parts of which are still perfect.
Most of the stones, however, have been used as building
material, and the church is said to be almost wholly
constructed of them. Traces of the guard-room may be
found near the north gate, though now luxuriant herbage
covers the ground, for the area of the station has long

been under the plough. To the passer-by it seems a level close of about eight acres, enclosed by a mouldering rampart shadowed with bramble and ancient thorn. Many of the lettered and sculptured stones are in the Cathedral Library. Of the Roman roads and bridges we shall speak in the section on Communications.

Roman Impluvium, South Shields

We should expect the Saxon period to yield far more remains of antiquarian interest than the earlier Roman period, but we are disappointed, for with the exception of crosses and grave-covers Anglo-Saxon relics are scanty. One however should be mentioned, a beautiful bluish-green glass bowl found in a Saxon burial-place at Castle Eden in 1775. We must remember that our

Saxon ancestors trusted little to walls and disliked confinement in buildings. Their settlements were strong in the boundary of wood and marsh rather than in earthwork or fortification. We know little, therefore, of the history of military or other buildings between Roman and Norman times. All the stone-work that can with certainty be described as Saxon is confined to churches. It was the activity and enthusiasm of Benedict Biscop, the teacher of the greater Bede, that gave the impetus to the building of stone churches; and with the two most famous, those at Wearmouth and at Jarrow, he was intimately connected. Deeming the wooden erections unmeet for God's worship he brought masons back from Gaul to erect a church in the Roman style which he had always admired. He sent for workers in glass; and from these French artisans the English learnt the art of glazing windows and making vessels of glass. The square tower and a portion of the west wall are the only parts now remaining of St Peter's, Monkwearmouth (founded 674); and of the sister monastery at Jarrow, founded seven years later, the present church contains but a very small portion. Of Bede and his work we shall speak later, but one fact, however, should be mentioned here: at Jarrow or at Wearmouth was written the most valuable manuscript now existing of the ancient Latin version of the Bible. This is the famous Codex Amiatinus with its beautiful writing and its gorgeous illuminations.

Saxon crosses and grave-covers are plentiful. Many of the most interesting are in the Dean and Chapter's Library; but one from Aycliffe churchyard is in the museum at Cambridge, and Bishop Tidfirth's stone, inscribed with his name in the old runic characters, was removed from Monkwearmouth to the British Museum. Two of the sculptured crosses yet stand where they were placed at Aycliffe. Bishop Ethelwold's cross, the most interesting of all, has passed through many adventures. Its fragments, now joined together and safely housed in the Library, were discovered, together with a number of medieval gravecovers, built into the tower of the ancient church of St Oswald near Elvet Bridge. The cross of which the fragments formed part was most likely the one brought from Lindisfarne and carried

Monkwearmouth Church Tower

about by the monks during their long wanderings with
the body of St Cuthbert. Leland, the antiquary, on
his visit to Durham saw it erect in the graveyard of
the cathedral, and he describes it as "a cross of a seven
foot long, that hath an inscription of diverse rowes in
it, but the scripture cannot be read. Some say that
this cross was brought out of the Holy Churchyard of
Lindisfarne Isle." In the Library is also St Cuthbert's
Cross, a beautifully wrought gold cross about three
inches long, having four equal arms and a large reddish
stone in the centre (see p. 3). When in 1827 St
Cuthbert's shrine was opened the cross was found under
three thicknesses of silk on the skeleton. The body of
Cuthbert now rests behind the altar-screen and there we
are shown the stone worn hollow by the knees of pilgrims
who had come to worship at his shrine, and who enriched
the cathedral by their offerings.

In the Galilee Chapel of the cathedral is the tomb of
the Venerable Bede. He was buried in his monastery at
Jarrow; but his body was stolen thence by an enthusiastic
monk of Durham, in order that it might be placed beside
that of St Cuthbert in the great building that was being
erected over his shrine.

The Cathedral Library contains a number of Anglo-
Saxon grave-covers collected by Canon Greenwell, some
of them of singular interest.

The manuscript of the Boldon Book, the Domesday
of the Palatinate, in the Library is a copy. The original
is supposed to have been lost during the time when the
great Cardinal Wolsey was bishop of Durham, though he

never once entered the county. He desired, it is said, to know what his revenues from the see were; and, when his possessions were confiscated by Henry VIII, the Boldon Book was included among them, and it cannot now be traced.

Bede's Tomb

19. Architecture—(*a*) Ecclesiastical.

A preliminary word on the various styles of English architecture is necessary before we consider the churches and other important buildings of our county.

Pre-Norman or, as it is usually, though with no great certainty termed, Saxon building in England, was the

work of early craftsmen with an imperfect knowledge
of stone construction, who commonly used rough rubble
walls, no buttresses, small semi-circular or triangular
arches, and square towers with what is termed "long-
and-short work" at the quoins or corners. It survives
almost solely in portions of small churches.

The Norman conquest started a widespread building
of massive churches and castles in the continental style
called Romanesque, which in England has got the name
of "Norman." They had walls of great thickness, semi-
circular vaults, round-headed doors and windows, and
massive square towers.

From 1150 to 1200 the building became lighter, the
arches pointed, and there was perfected the science of
vaulting, by which the weight is brought upon piers and
buttresses. This method of building, the "Gothic,"
originated from the endeavour to cover the widest and
loftiest areas with the greatest economy of stone. The
first English Gothic, called "Early English," from about
1180 to 1250, is characterised by slender piers (commonly
of marble), lofty pointed vaults, and long, narrow, lancet-
headed windows. After 1250 the windows became
broader, divided up, and ornamented by patterns of
tracery, while in the vault the ribs were multiplied. The
greatest elegance of English Gothic was reached from
1260 to 1290, at which date English sculpture was at
its highest, and art in painting, coloured glass making,
and general craftsmanship at its zenith.

After 1300 the structure of stone buildings began to
be overlaid with ornament, the window tracery and vault

ribs were of intricate patterns, the pinnacles and spires loaded with crocket and ornament. This later style is known as "Decorated," and came to an end with the Black Death, which stopped all building for a time.

With the changed conditions of life the type of building changed. With curious uniformity and quickness the style called "Perpendicular"—which is unknown abroad—developed after 1360 in all parts of England and lasted with scarcely any change up to 1520. As its name implies, it is characterised by the perpendicular arrangement of the tracery and panels on walls and in windows, and it is also distinguished by the flattened arches and the square arrangement of the mouldings over them, by the elaborate vault-traceries (especially fan-vaulting), and by the use of flat roofs and towers without spires.

The medieval styles in England ended with the dissolution of the monasteries (1530–1540), for the Reformation checked the building of churches. There succeeded the building of manor-houses, in which the style called "Tudor" arose—distinguished by flat-headed windows, level ceilings, and panelled rooms. The ornaments of classic style were introduced under the influences of Renaissance sculpture and distinguish the "Jacobean" style, so called after James I. About this time the professional architect arose. Hitherto, building had been entirely in the hands of the builder and the craftsman.

We may now turn to our county, which, as befits the dominion of a priest-king, is so rich in noteworthy buildings erected for worship that we cannot here mention more than a tithe of them.

The rude work of the Saxon builders survives nowhere as a complete building, though much of the little church at Escomb, near Auckland, and the blackened, beaten towers of the twin churches at Monkwearmouth and Jarrow, are Saxon. With the Normans came a time of strenuous activity in the erection of churches, castles, and beautiful abbeys; and in Durham the vast revenues of the bishopric and the great gifts at St Cuthbert's shrine must have been largely spent on buildings. The crypt in Durham Castle is our oldest Norman relic; but the grandest of all, the noblest work in the world of the Norman builders, is the portion of the cathedral comprising the nave and transepts. The Norman style, as we have seen, prevailed from the Conquest until about 1200; and when we consider the aggregate of the cathedrals, castles, and abbeys erected in our country during that short period, we are amazed. The total far exceeds the mass of public buildings built in any country whatever during a like period. And nowhere was building activity so great as in Durham.

Of Early English work we have splendid examples in the Chapel of the Nine Altars in the cathedral, the churches of Hartlepool, Darlington, Billingham, and the chapel at Auckland, perhaps the finest collection of Early English buildings in the country. The Chapel of the Nine Altars is indeed the crowning glory of the cathedral. This light and elegant style of building with its high pointed arches struck from two centres was probably invented in this country. It was certainly here brought to perfection; and the traditions of other countries ascribe their most beautiful churches in this style to English artists.

The Nave, Durham Cathedral

The Early English style passed gradually into the Decorated, and it is a striking fact that there is not in

the county a building of any note in either this or the
Perpendicular style which followed.   Even to-day there
is a remarkable scarcity of spires on our churches :
Darlington, Conscliffe, Chester-le-Street, Boldon, and
Ryton are the only old ones.   The harassing wars between England and Scotland and the active part taken
in the struggle by the Princes palatine and the men of
Durham were quite fatal to building operations.   There
was of course some work effected in the prevailing style
in spite of wars.   Windows at Houghton-le-Spring and
elsewhere, alterations at Finchale Abbey, and, finest of
all, the wonderfully carved altar-screen and the bishop's
throne in the cathedral, are examples of Decorated architecture.   And during the prevalence of the Perpendicular
style, Bishop Langley—who was a Cardinal of the Church
as well as a Prince in the State—laid out large sums of
money for building purposes.   This, however, was devoted
to necessary repairs of earlier foundations, so that the
Perpendicular features are simply additions to buildings
mainly of another style, e.g. the large west windows of
the Galilee Chapel.

The cathedral, though in the main a Norman building,
contains examples of all these styles.   The genius of many
architects over a long stretch of years and the great natural
beauty of its site, have combined to make Durham unique
in beauty and grandeur among the cathedrals of England.
Distant views of the exterior, especially that from Pre-
bend's Bridge, can nowhere be surpassed for impressive
beauty ; and the interior, from the ponderous sanctuary
knocker to the shattered sculptures in the crypt, is full

of interest. The gigantic nave terminating in the wonderful rose window in the east, with its huge rounded columns and massive arches, together with its attendant aisles, is the most perfect specimen of Norman architecture in existence. Its grand simplicity is relieved from monotony by the variety in the pillars, some huge cylinders with small fluting, others clusters of round

The Galilee Chapel

shafts, others adorned by bold zigzag carving. This variety is imitated in other churches that owe their origin to the bishop's fund: at Darlington, Hartlepool, Auckland, and, still more strikingly, in the little old church at Pittington.

The famous Galilee Chapel of Bishop Pudsey at the west end is very late Norman, bordering upon Early English, and yet it is unlike either style; the repetition of the arches and their sumptuous decoration giving almost the impression of a Moorish building. The piers of the arches are clusters of four slender shafts : two of the shafts are of Purbeck marble, for Bishop Pudsey, lavish in all he did, brought "sundry pillars of marble stone from beyond the sea" (i.e. from Dorset) to adorn his chapel ; the other shafts are of sandstone quarried quite near, and were used to give an appearance of stability otherwise lacking, though the whole weight of the arches is actually carried by the original pillars.

Behind the choir is the Chapel of the Nine Altars, one altar under each of the windows. Two of these altars are the shrines of Cuthbert and of Bede. The chapel, the largest attached chapel in the kingdom, is probably the finest surviving example of the Early English style, and the wonderful carving of the capitals can nowhere be equalled. But the original narrow lancet windows have been replaced by magnificent specimens of Decorated and Perpendicular style. The most noteworthy perhaps is the double-ribbed and elaborately decorated north window.

Two of the internal structures are of remarkable beauty—the Bishop's throne and the altar-screen. The first shows great richness of chiselling combined with simple grandeur of design : it was built by Bishop Hatfield over the vault prepared for his body in order "to inculcate upon his own heart a lesson of humility under the almost regal distinction to which he had risen." The double-fronted altar-screen was made in London by the most skilled carvers of the day and

St Cuthbert's Church, Darlington

was erected in the cathedral in 1380. It is of Caen
stone and is the most elaborate example in the world of
canopied and pierced screen-work.

The church of St Cuthbert, Darlington, is perhaps
the grandest of our parish churches. It was built about
1230 in Early English style; but a massive stone gallery,
the whole width of the chancel arch, is of the Decorated
period. The stone screen was probably added to support
the weight of the central tower with its fourteenth century
spire—one of the very few in the county. St Andrews,
Auckland, founded by Bishop Bek about 1300, is a
complete Early English church in all its main features—
except the roof, a flat timbered one of Tudor times. It is
the largest parish church in the diocese. The church at
Conscliffe, also of this period, is built in an extraordinary
position; overlooking the Tees from the summit of a limestone steep that has been cut into a high perpendicular
wall. Staindrop church, the nave of which is late Norman (about 1200), the other parts more recent, was the
burial place of the Neville family, the lords of Raby
Castle, and contains their elaborate altar-tombs.

Chester-le-Street, where Bishop Bek founded a great
collegiate church, has a tapering spire 156 feet high on a
tower half square, half octagonal—a curious design. To
the top of the square tower the church is Early English,
the octagon tower and the spire are late Decorated. It
contains what has been described as "an aisle of tombs,"
effigies of the lords of Lumley, which show the costumes
of five centuries. One may see from Chester church how
well the ancient architects suited their designs to their

sites.   Its lofty spire is applicable to the protected valley;
the stunted spire of Boldon church is well fitted to en-
counter the gales that beat on its exposed situation; and
at Medomsley, on the summit of the unsheltered hills
between Durham and Northumberland, we have an Early

Merrington Church

English church with a squat, flat-roofed nave, a chancel and
vestry, all lower than the modern cottages that adjoin it.

The strong square tower of Merrington church is
another instance of suitability.   The church stands on
an exposed hill overlooking the Wear Valley past Durham

9—2

and Chester. It is, indeed, as many of the Durham churches were obliged to be, more of a castle than a church; and in 1140 it actually sustained a siege, being defended by the forces of the usurping bishop, Cumin, and attacked by those of the duly elected one, William de St Barbara.

The small church of Brancepeth is remarkable in many ways : its fine tower is Early English, the eight-sided shafts and the arches of the nave belong to the Decorated period, and the chancel is partly Perpendicular, as is the fine choir screen. The fittings of the church are Tudor and are splendid specimens of carving. In the chancel are the effigies of some of the Nevilles, lords of Brancepeth as well as of Raby.

St Mary's church, Gateshead, is of the late Perpendicular period. Of the original Norman building—where Bishop Walcher was killed—the doorway on the south side only remains.

Here and there throughout Durham are scanty remains of once great monasteries or nunneries. The followers of St Benedict were most numerous in Durham, but there were also establishments of Dominicans; and there was one, at Baxterwood Priory near Durham, of Austin Canons.

In Saxon times there were monasteries at Wearmouth and at Jarrow, or rather these two places were branches of the one monastery; there was also St Hilda's monastery at South Shields, but all trace of this has long disappeared. William of Carileph, the builder of the cathedral, who greatly disapproved of the worldly lives led by the Saxon monks at St Cuthbert's shrine, dismissed them and founded at Durham a college of Benedictine monks. This was St Cuthbert's Priory, of which a branch (or cell) was Finchale Priory, and another the "lovely retreat," Beaurepaire, now the colliery district of Bearpark. At Neasham, two miles east of Barnard Castle, there was a Benedictine nunnery; and at Barnard Castle, itself an establishment of Austin Friars, there was another.

Finchale Abbey

The ruins of Finchale Abbey, the most extensive in the county, are in a beautiful dell on the south bank of the Wear, where it was sheltered by the woods and rocky crags of Cocken on the opposite bank. Little now remains of the once great building except some half-ruined Early English columns and windows, and some Perpendicular work.

Close to Hartlepool church is a house known as The Friary; and we know that in early days there existed both Grey Friars and Friar Preachers at Hartlepool. But most of the building, which is now converted into the union workhouse, is evidently of late date. This particular monastery was founded in 1258 by Robert the Bruce.

Hospitals, or almshouses as we should now call them, were founded by charitable men as branches of these various religious houses; and some in Durham still remain, though now greatly changed. Sherburn Hospital, quite close to Durham city, was built by Bishop Pudsey in 1181 for the reception of lepers; but since that disease has died out of our country it has been converted into an almshouse. The present building was erected after 1300, for in that year the Scots burnt the original edifice. Greatham Hospital near Stockton was built from the Durham estates forfeited to the bishop from Simon de Montfort after his defeat at Evesham.

Of Kepier Hospital, founded by the famous Bishop Ralph Flambard in 1112, nothing now remains but the gateway, a triple archway flanked by the apartments of the porters. The Early English pointed arch indicates that the building is much later than the first establishing of the charity.

20. Architecture—(*b*) **Military.**

We should expect Durham, long the frontier of Roman civilisation and longer the bulwark of England against the Scots, to be rich in fortified buildings. And this is what we find.   The Normans especially, often dwelling in small bands among a hostile population, developed the building of castles into a science.   Against the frequent inroads of marauders from Scotland and from the moors of Northumberland even the old farmsteads were built in the form of a keep into which the cattle might be driven when alarm was given.   After the union of the kingdoms and the consequent cessation of border warfare, there was no longer need for buildings of which the main purpose was protection.   And in several of the Durham castles that yet remain, alterations have made the domestic more prominent than the military element.   Some of the oldest castles, like that of the Bruces at Castle Eden Dene, have so entirely vanished that even their site cannot with certainty be indicated : the present representative is an entirely modern building.

The earliest fortresses built by the Normans were lofty stone keeps like that which still confronts us as we pass from Gateshead over the High Level into Newcastle. The strength of these keeps depended on the thickness of their walls and the natural advantages of the position. They were usually three stories high; the lowest contained a well—often of great depth—and the store rooms; the middle provided the soldiers' quarters; and the third was the home of the governor of the castle and his family. Later one, and then two outer walls enclosed a space around the keep, an inner bailey, and an outer bailey. If the defenders were driven from the outer wall the inner was a second line of defence, and the keep was their last resort.   A moat crossed by a drawbridge usually surrounded the castle and the entrance gate was defended by a portcullis that could be dropped at will. The outer defences of castles like Raby and Brancepeth enclosed a space of several acres.

Durham Castle, built by William the Conqueror in 1072, in order to keep his unruly northern subjects under control and to afford protection to the bishop whom he

Durham Castle

had appointed, is the oldest of all the Norman buildings in our county. Parts of the original structure still remain, but most of the present castle is of a later date. It is the finest example we have in England of a mount-and-bailey fortress on a strong triangular site. On the north and west are precipitous natural defences; the guarded entrance was on the one accessible side from the platform

on which the cathedral and monastery stood.   The two
ancient bridges, Elvet and Framwellgate, carry the north
road over the Wear to cross the neck of the peninsula on
which the castle stands.   Within the outer and inner
baileys was the shell-keep, on the summit of the rock.
Outside the road a moat between the bridges could, it is

Raby Castle

said, isolate the old town; and Claypath, the street leading
from the Market-place towards the moor, began at the
gate of the sluice (the *Cleur-port*).   King William's
chapel in the crypt is the oldest part; the tall cylindrical
shafts with their curious capitals and the lofty vaulted roof
are typical of the earliest Norman buildings.   The most
interesting and beautiful parts of the fortress are attributed

to Bishop Pudsey: the Norman doorway in the lower gallery may fairly be called the finest example of late Norman art in England. The "Great Hall," attributed to Bishop Hatfield, is now the dining room of Durham University.

Raby Castle, on a gentle slope north of Langley Beck, is the most perfect of our castles and is yet occupied, with not a great deal of alteration, by the descendants of Milton's Sir Henry Vane, "Vane, young in years, but in sage counsel old." Its embattled walls and its entrance tower, defended by a portcullis and side parapets, still remain entire. The castle, which is rectangular, was built round a main courtyard and was formerly surrounded by a moat. Two towers defend the corners, one, Bulmer's Tower, of unusual lozenge shape. Before the "Rising of the North" it was the seat of the powerful house of Neville, "Seven hundred Knights, retainers all Of Neville, at their Master's call, Had sate together in Raby's Hall." During the struggle between King and Parliament it was twice besieged and once taken by the royalists, and once besieged and retaken by Cromwell's forces.

Barnard Castle is now in ruins, but it must once have been the most important and extensive fortress in the north. Bernard Baliol, ancestor of Robert Bruce's rival, founded the castle on the precipitous rocks that overhang the Tees, and called it after his own name, and the town of Barnard Castle grew up under the shadow of his fortress. The outer ward or bailey enclosed seven acres of ground ; the inner ward and the round keep were built at a distant angle of the enclosure, where the steep outer slopes and the swiftly-flowing river at the foot afforded protection. John Baliol, one of its lords, founded Balliol College at Oxford, and it was the son of the Oxford founder whom Edward I chose as King of Scotland in 1292. The castle sustained a siege of eleven days against the whole of the rebel forces during

Ravensworth Castle

the "Rising of the North" (1569); and its defender, Sir George Bowes, surrendered only when his garrison were deprived of water, of which the besiegers had cut off every supply.

Lumley Castle stands on an elevation above the Wear, a mile east of Chester-le-Street. It is a large manor house rather than a fortress. The building is a quadrangle with an enclosed area, and at each angle are projecting octagonal turrets which give an unusual appearance to the castle. The building material is yellow freestone quarried on Gateshead Fell.

Ravensworth Castle is built on a beautifully wooded slope rising from the Team, a mile or two from its junction with the Tyne. It must have been a fortress

Brancepeth Castle

before any records now in existence were made. There is no licence to embattle this house in the bishop's documents, but of every other castle in the county we have that record. The present building, however, is not the original, but a later edifice of Early English style.

Close to Durham is Brancepeth Castle, finely situated on a wooded slope rising from the north bank of the

Wear, though the collieries are now pressing close upon it. It is another example of an ancient castle adapted for a modern mansion. Two bold, irregular towers crown the west and south sides and these are said to be of ancient construction ; the rest of the building is quite modern. At Brancepeth Castle the northern gentry and their retainers met for the " Rising of the North," which brought about the outlawry of Neville, the owner of the castle.

Hylton Castle, quite near Sunderland, appears to have consisted of a great keep alone. Both for its size and decoration it is remarkable, but no part is of great age. Square piers crowned with eight-sided turrets line the front of the castle and divide it into compartments.

Many of the old fortresses have been transformed. More, however, have vanished, overwhelmed by the growth of industry and commerce, and are to be traced only by a few records or relics. That which gave its name to Castle Eden has already been mentioned ; another is Stockton Castle, once celebrated as a Norman fortress and garrisoned by the king's troops during the Civil War. It was long a seat of the bishops of Durham, " that had a great moat about it." Now it has vanished and its stones are built into the houses of Stockton.

21. Architecture—(c) Domestic.

The wealth of ecclesiastical and of military buildings
is in great contrast to the paucity of notable mansions
in the county. This is not to be wondered at, for in
earlier times the distance from Court, and in later times
the industrial character of the county, made Durham not
very attractive for the seats of the nobility. Not one of
the great mansions scattered up and down the county is
in its entirety older than Elizabeth's reign.

The oldest houses are either transformed castles or
buildings adapted for defence. The ruins of Langley
Hall, once the seat of the Scrope family, are strongly
posted on a hill overlooking Durham ; at Holmside, two
miles from Lanchester, are the remains of an ancient
moated house ; and at Butterby we may see the old
gate-house and the moat of another protected dwelling.
In the oldest parts of the towns we also note how defence
was thought of before comfort. In Hartlepool the great
stone walls and the ancient gateways still remind us of
days when the raids of the Scots were dreaded, and in
Middlegate Street the overhanging upper storeys of many
of the houses tell a like tale.

Among the oldest buildings is Gainford Hall, near the
Roman station at Piercebridge. It is a picturesque many-
gabled building with square mullioned windows like those
of Horden Hall, which we noted during our survey of
the coast. Barnard Castle contains some very old houses :
one is the curious Blagroves House where Oliver Cromwell,
in spite of his battering the castle, was regaled with cakes

and burnt ale. The rectory at Ryton is a small gabled house mainly built about 1700, but some parts are Elizabethan. The old workhouse in Durham market-place, too, is probably Tudor.

Old House on Elvet Bridge

When Bishop Van Mildert gave up his castle at Durham to be the home of the Northern University, the Manor Place at Auckland became the Palace of the see. It occupies a delightful situation in a wooded park at the confluence of the Wear and the Gaunless. Though built originally by Bishop Bek the present mansion is fairly modern, rebuilt almost entirely by Bishop Cosin after the Restoration had once more brought him back to his diocese. The private chapel in Early English and Perpendicular styles, with its magnificent wooden screen, is the most notable part of the house.

Stella Hall near Ryton is a magnificent old gabled house close to the Tyne: it was the seat of the Tempests, and Lord Widdrington was deprived of it on account of his complicity in the rebellion of 1745. Gibside Hall, a long low building of two storeys, stands not far from Gateshead in the midst of delightful grounds. It was the home of John Bowes, whose wife built the magnificent Bowes Museum at Barnard Castle. The pretty moorland village of Bowes, from which the family name of the active, restless Sir George Bowes is derived, stands near Barnard Castle. Here is the original "Dotheboys Hall" of *Nicholas Nickleby*. The old pump, where Dickens makes the pupils of Mr Squeers wash themselves, still stands in the yard.

Whitworth Hall near Spennymoor, the seat of the Shafto family, is of recent date, and Mainsforth Hall, close to Ferryhill, where Robert Surtees spent over thirty years in preparing his great *History of Durham* is of no great age. Many fine mansions have been built during the last century in Durham, as a result mainly of its rapid growth in wealth, the finest being Wynyard House, the home of the Marquis of Londonderry, a splendid specimen of modern building which stands in an extensive park north of Stockton.

Even to-day, when carriage is so easy and cheap, nearly all the houses in Teesdale beyond Barnard Castle are built of stone, and the roofs are often of the same material. But in the industrial parts the houses are of brick. In too many hastily-built villages there are long dismal rows of cottages straggling over the ruined fields around the collieries or the ironworks. A great change, however, is now in operation, and some of the newer villages are much better. In many instances the workmen purchase, in some cases indeed they build in their spare time, the houses in which they live. At Jarrow, for example, practically the whole town belongs to the wage-earners ; and this is the case also at Dunston-on-Tyne, where the houses are rapidly climbing up the hill overlooking the Tyne and the double city, Newcastle and Gateshead. Building societies are nowhere more popular. The chief material is still clay-brick, but modern sanitary arrangements are generally found, and the houses often have a little strip of garden.

22. Communications: Past and Present.

The earliest roads in our county were no doubt the trails that guided our remote ancestors through the forest or over the moorland. Perhaps even these were originally the tracks which the wild animals had made. The best and most plainly marked were not better than the rough footpaths by which even to this day we make our way across the heathery, boulder-strewn moors on the west, where scarcely legible guide-posts alone enable the stranger to find his way.

Throughout our history from Roman times, however, Durham has been on the main line of communication. The county was part of the route over which the Roman legions marched to beat back the attacks of barbarians from beyond the wall; expeditions from Scotland to England and from England to Scotland with rare exceptions chose the east coast road along the flanks of the Pennines; and to-day the North-Eastern Railway is a link in the line which along the Great North Road provides the most rapid communication between London and Edinburgh.

Long before St Cuthbert's shrine had made Durham rich and famous the district was one of special interest and importance. During the Roman occupation—for a period, that is, longer than our rule in India has yet lasted—it was on the outskirts of civilisation. Here, at *Vinovium* (now Binchester) or *Pons Aelii* (now Gateshead) or Chester-le-Street or Ebchester, were posted the veteran soldiers who were to defend the Empire against the savage men of the north, and who in the intervals of fighting constructed the great trunk roads that bound the whole Empire closely together. Just north of Durham is the Tyne Gap through the Pennines, a stretch from sea to sea no point of which reaches 600 feet in altitude. In our day the railway between Newcastle and Carlisle finds easy passage through this gap, and here in Roman times was the road and the great wall that marked the farthest limit of Roman rule.

Little wonder that these Roman roads are not even yet destroyed. They were built regardless of cost in labour and time. Long after the decline of Roman power they remained visible signs of the influence exerted on the world by that wonderful people. Into Durham a branch of the main road, Watling Street, entered from the south at Piercebridge and struck almost due north across the higher ground to Binchester. Newgate, the main thoroughfare of Bishop Auckland, was one part of this "street." Lanchester was probably the chief camp along the road. At Ebchester the bridge over the Derwent carries it into Northumberland. Another branch goes through the camp, Chester-le-Street, to the "head of the street," Gateshead. Here it was that the Emperor Hadrian built the Pons Aelii across the Tyne, an event considered of such interest that at Rome a medal was struck commemorating it.

The modern representatives of the old Roman roads are still under the care of the public authority. To the Durham County Council a grant in aid is made by Government for the repair and improvement of the great trunk roads. There is in the county an abundance of good road-metal, and the roads and public bridges are admirably kept.

With the nineteenth century and the vast amount of trade that followed on the use of steam in industry, the need for rapid and easy means of communication became urgent. The traffic up the three rivers was great, but the streams were then undredged and shallow, and the journey was terribly slow. To sail the ten miles up the

Tees to Stockton took as long as to sail from the Thames to the river. Just as later the congestion of traffic on their canal prompted the merchants of Liverpool to send for George Stephenson to build the Liverpool and Manchester Railway, so the difficulty of navigating the Tees was the origin of the Stockton and Darlington Railway, the forerunner of the passenger lines that now encircle the globe.

A canal, indeed, had been suggested to join Darlington with the Tees at Stockton, but the project came to nothing. The Tees Navigation had made the New Cut, the nearest approach to a canal in our county, and so saved a mile or two. But something much more speedy than canal navigation was wanted, and in 1822 the first rail of the first railway along which locomotives drawing carriages with passengers and goods were to run was laid with great ceremony at Stockton.

The work of George Stephenson is mentioned elsewhere : he was not indeed the first to realise that the power of steam might be applied to a moving as well as to a stationary engine, but it was he who made the locomotive a practical success and initiated the marvellous system of iron roads that we know to-day. On September 27th, 1825, the first train ran, drawn by the engine, Locomotion No. 1, now standing on a pedestal of honour in Darlington Station. One idea of Stephenson, in a quite accidental manner, has linked his name with our railways in a way that is now unalterable; he adopted for the line the width between the rails of which, as being best for the colliery carts drawn by horses along the lines, he had had experience at Wylam and Killingworth. This width, 4 feet 8½ inches, afterwards successful in its competition with the "broad gauge" of 7 feet, is now the gauge of all the main lines of Great Britain. Locomotion No. 1 in 1825 rumbled from Darlington preceded by a man carrying a red flag; to-day the North-Eastern runs from Darlington the fastest train in the British Isles. The express that leaves

"Locomotion No. 1"

for York at 1.9 p.m. covers the 44¼ miles in 43 minutes, at the rate of 61·7 miles per hour. Another remarkable contrast and illustration of the rapid development of railway engineering is afforded by comparing the oldest iron railway-bridge, that which carried the pioneer line over the Gaunless at West Auckland, with the massive bridge at Sunderland completed in 1909, where the huge span of steel in the middle, the heaviest in Britain, rests

on massive granite piers 300 feet apart. As in the other
and earlier railway bridge, the High Level between
Gateshead and Newcastle, the railway runs above the
roadway.

Except for a few short colliery and works lines, the
North-Eastern Company has the monopoly of rail communication in our county. It forms a link in the East

King Edward Bridge, Newcastle

Coast Route—the old North Road—between London
and Edinburgh. The course of the main line crossing
the Tees from Yorkshire is almost due north. It runs
along the river Skerne into Darlington, to which busy
centre the locomotive works for the line have recently
been transferred. From Darlington it crosses the mining
and iron-working region to Durham. Hence to Gateshead
is another crowded mining and industrial district, and as

we near the Tyne the dirt and smoke become even more conspicuous. Into Newcastle the expresses go by the new King Edward Bridge, opened for traffic in 1906.

From the main line spurs are thrown out to the west up the dales till the lack of population makes further railway building unprofitable. In Teesdale, Middleton is the terminal; in Weardale, Wear Head. West of these places roads and rough tracks cross the mountain ridge to Alston or Appleby. To the east loops are thrown down the valleys till the coast towns are reached. From Darlington to Hartlepool and looping back to Auckland and Durham, from Durham to Sunderland looping again to Newcastle, and in between, as the colliery maps show us, is a very network of lines connecting the mines with the seaports and with the works, carrying imported ore to the furnaces or machinery to the ships, serving as handmaid to the multitude of industries that depend on the mineral wealth of the county. For the North-Eastern from the first has found its chief profit in minerals and merchandise rather than in passengers, and it has been zealous in developing trade, home and foreign, because of this. It is, for instance, the greatest dock-owning railroad, and in our county owns the docks at Dunston, Tyne Dock, and Hartlepool, though not at Sunderland and Seaham.

The bridges of our county, old and new, are important and interesting. We notice that *ford*, frequent in the names of towns elsewhere, is almost absent here—Gainford on the Tees is one of a very few. The steep banks and rapid course of the upper streams rendered foot passage so dangerous that bridges were a necessity. Early in the reign of the bishops the road across the neck of land joining Durham to the country around was carried by two fine bridges over the Wear—Elvet and Framwellgate. It is said that for security in times when a Scottish raid was ever a possibility a moat that could at will be flooded cut off the cathedral from the bridges. The gate of the

Old Knitsley Viaduct

sluice was the *Cleur-port* and the French word as already mentioned survives to-day in Claypath, one ot the chiet streets of Durham. Some of the railway bridges have already been alluded to : another interesting one was the wooden trestle viaduct at Knitsley on the Durham and Blackhill loop. It carried the railway over Smallhope Burn, but the valley it crossed is now filled in and the trains cross the valley on a solid causeway.

23. Administration and Divisions— Ancient and Modern.

About the organisation of the British dwellers in our county—the Brigantes, "the people of the heights," one of their chief tribes—we know very little. But we know a good deal about that of the Angles who gradually formed their settlements here in Deira. The unit group or community was a "township," corresponding to our parish. They had their dwellings within an enclosure, or *tun*, that protected them from their enemies and their herds from wild beasts. Hence we have the numerous names ending in *ton* or *don*. The men of the township met generally at some conspicuous natural object such as a big stone or tree, or at some other chosen spot, in order to discuss their affairs and make their *by-laws* (laws of the *by* or township). All adult males might attend and take part in the deliberations. It is a most interesting fact that not many years ago a partial return was made to this primitive method of settling public questions by a meeting of all concerned. By an Act of Parliament of 1894 Parish Meetings were established for places with fewer than 300 inhabitants. To-day, therefore, all men in such places may meet in their schoolroom, as their forefathers met long ago in the open field, to manage local affairs.

The units were further grouped for administrative purposes into "hundreds" or "wapentakes," or, in our county, into "wards." The hundred may have been an area containing the land originally settled by a hundred families, or simply a hundred hides of land—a hide was a portion containing from 60 to 120 acres. Most likely, however, the name signified the district that was called on in time of need to furnish a hundred warriors to the host. Both wapentake and ward are military terms, and suggest that, when danger threatened, an organisation existed for defence. Durham was in former days divided into four wards, named after the meeting-place Darlington, Stockton, Easington, and Chester—three *tuns* and the old camp city. The position of the places, all in the east coastal plain, makes clear to us that the early settlements were limited to the seaboard. As the land rose towards the hills population ceased, or consisted only of unorganised bands of the earlier inhabitants. Later, as the interior became more thickly settled, a fifth ward, that of Durham, was constituted. The officials of these wards were all answerable, not to the King in London, but to the Bishop at Durham. But there was in later days within the county a " liberty " or " honour," that is, a district under its own lord and exempt from the authority of the Bishop. The liberty of Gainford and Middleton was granted by William Rufus to Guy Baliol, the ancestor of the claimant to the Scottish throne. His son, Bernard, founded his castle on the high rocks overhanging the Tees, and Barnard Castle became the name of the liberty.

The meeting—or moot—of the whole shire was a body of great importance. It met twice a year for the transaction of business—business which usually consisted in deciding how much money should be raised to meet the Bishop's needs, and how that money was to be raised. It was attended by twelve elected men from each ward, and by four men with a *reeve* as leader chosen from every township. The former may be regarded as similar to the Aldermen in our present-day County Council, the latter to the ordinary Councillors. Much later a King's officer, the reeve (steward) of the shire, or *Sheriff*, was appointed to look after the King's interests.

But in a thousand years our population has increased forty-fold, and government is no longer concerned simply with raising men and money for war and with the keeping of order. It has become much more complicated. Townshipmoots, Wardmoots, and Shiremoots have expanded into more regularly constituted Parish Councils, District Councils, and County Councils. And instead of twice a year the County Council meets four times, and its many varied Committees are constantly working. For all these governing bodies the members are elected by the people, so that we all have power to influence the ruling of our localities. It is a matter for regret that more people do not take an active interest in the affairs of their district.

The astonishing growth of our large towns during the last two centuries has led to the creation of a new area of local government. Many towns have such a number of inhabitants that a separate and independent Council is required for them. These large towns have, therefore, been made into County Boroughs. Their Town Councils have all the powers of the great Council of the County, the Lord Mayor or the Mayor in each

case performing the duties of the Sheriff. In Durham
there are four County Boroughs : Sunderland, Gateshead,
South Shields, and West Hartlepool. Within these the
County Council has no authority.

Other towns there are, not quite populous enough to

The Town Hall, Stockton-on-Tees

be created county boroughs, but sufficiently important to
have for most purposes a governing body of their own. The
County Council has a varying amount of authority over
them, though most questions are settled by the Mayor
and his Town Council. The Municipal Boroughs, as

these towns are called, in Durham are, in their order of
population according to the census of 1911, Darlington,
Stockton, Jarrow, Hartlepool, and Durham.

The meetings of the County Council are held four
times a year in the Shire Hall at Durham.   When it
assembles at its quarterly meetings, it considers and

Municipal Buildings, South Shields

discusses reports, and authorises plans suggested by its
Committees.   These smaller groups are the active working bodies, and each Councillor belongs to one or more
of them.   One Committee, for instance, looks after the
immense County Asylum at Sedgefield.   Then we have
a most important Committee that looks after the health
of the county, and others that devise means for the

prevention of diseases in animals, for the organisation of the education of the people, etc., and as the raising and spending of money always follows on the action of government, there is the very important Finance Committee. In the County Boroughs and Municipal Boroughs the work of the governing bodies is on much the same lines.

The Durham County Council contains 78 Councillors elected by the ratepayers, and 26 Aldermen elected by the Councillors. The Councillors serve for three years, the Aldermen for six years; in other respects they have the same powers.

Every holder of the franchise has to do with two other governing bodies, the Guardians of the Poor and the great assembly in Parliament. The district attached to the parish church was once the unit for the care of the poor, but later most parishes were joined together into larger areas called Poor Law Unions. There are 16 Members of Parliament chosen in the county, eight corresponding to the old Knights of the Shire from divisions of the county outside the large towns, and eight from the boroughs corresponding to the Burgesses first summoned by Simon de Montfort, two from each borough. Sunderland is now the only borough that returns two members; the other six—Gateshead, South Shields, The Hartlepools, Stockton, Darlington, and Durham City—return one each. Of these Parliamentary Boroughs, Durham City contains a less number of voters than any other constituency in England. The county division of Chester-le-Street had 25,102 voters in 1912, Durham City had only 2601:

the member for Chester-le-Street thus represents nearly ten times as many voters as the member for Durham City.

We find, rather strangely, that as the government of our country became more complicated, instead of existing areas being taken for the new purposes, quite new areas were devised. The areas for the choosing of M.P.'s, those for the purposes of the Poor Law, and those for the County Council Act of 1888, are often independent one of another. The first form the Ancient or Geographical County, the second the Registration County, and the third the Administrative County. In one case only— the neighbouring county of Cumberland—do the three areas coincide. Within the county we have such peculiarities as these: the Parliamentary Borough of "The Hartlepools" comprises the County Borough of West Hartlepool, part of the Municipal Borough of Hartlepool, and part of the Rural District of Hartlepool ; and the Urban District of Spennymoor extends into three Poor Law Unions.

24.   The Roll of Honour of the County.

The County of Durham, so long a kingdom in itself, has been the home of such a number of men of mark that a selection of its greatest men is a difficult task, and omissions must of necessity be many. The famous churchmen, both in early and in late times, and the men who by their energy and skill have made the county pre-eminent in industry and commerce, are particularly to be noticed. Great warriors too like the Nevilles, great statesmen like Lord Durham—the founder of our present colonial system—and great writers like Butler of the *Analogy* and Paley of the *Evidences* are no less conspicuous, though the two latter belong only to the county by residence, not birth.

Aidan, the missionary bishop; Cuthbert, who won such ascendancy and fame as no churchman north of the Humber has rivalled; Hilda, who for eight years presided over the oldest nunnery at Hartlepool till she migrated to the more famous abbey of Whitby; Bede, the historian and scholar; Bury, the munificent patron of learning; Pudsey, the mighty builder; the regal and magnificent Bishop Bek; these are but a few of the leaders who have adorned the northern church. Into his twin monasteries of St Peter at Wearmouth and St Paul at Jarrow, Benedict Biscop collected all the best learning and all the best art of his time; and for a long while the central light of Christendom was between the banks of Wear and Tyne. Bede lived and worked in this light. He is the earliest and in many ways among the greatest of our historians, and he it was who first turned part of the Bible into a language that could be understood by the common people. Familiar to all of us is the affecting story of his anxiety, when he felt death at hand, to complete his translation of St John's Gospel, and of his resigned and joyful echo of the scribe's words, "It is finished." His body now rests under a plain slab in the Galilee, but his best memorial is the number of institutions for the promotion of learning called after his name.

Bede's *Church History of the English People*, which he completed in 731 in the peaceful time before the coming of the Danes, is our great authority for Saxon times in England. Later historians are Simeon of Durham and in the nineteenth century John Lingard. The latter, a native of Winchester, was head of the Roman Catholic college before its establishment at Ushaw Moor. There, a few miles from Durham City and forming one of the colleges of its University, is now the representative of the old Jesuit college of Douai.

In later days Robert Surtees of Mainsforth, the friend of Sir Walter Scott, wrote the *History of Durham* in four princely folios, a work inferior to none of our great county histories. Surtees lived a most retired life at Mainsforth, where he spent the best part of thirty years in the preparation of his book—which has been largely laid under contribution for this present volume.

The spread of learning has, indeed, long been a special work in our county. William of Durham—usually identified with William de Laneham, archdeacon of Durham and rector of Bishopwearmouth—in 1249 bequeathed money to found University College at Oxford; this is the earliest of our existing colleges, though Merton, founded in 1264, claims to be the first real example of the collegiate system. William of Durham's example was quickly followed by Hugh of Balsham, Bishop of Ely, who founded St Peter's College, Cambridge, in 1284. In the nineteenth century, in 1832, the last of the Prince-bishops nobly carried on the work of the great churchmen of old who had done so much to promote learning. In that year Bishop Van Mildert gave up Durham Castle to be the seat of the newly-founded Durham University. Another who should be remembered in this connection is Bernard Gilpin, the "Apostle of the North," who, though born in Westmorland, was for many years rector of Easington and then of Houghton. During Elizabeth's reign he founded at the latter place the famous Kepier Grammar School.

Bishop Bury, the friend of Petrarch and an ardent lover of books—the praise of which and the care that should be taken of them he celebrated in *Philobiblon*—established the first lending library in England as long ago as 1340. He was both a scholar, the most learned man of his country and age, and a patron of scholars, and he destined his choice library for the use of poor scholars at Oxford. Thus as the munificence of William of Durham was responsible for the first Oxford college, so the library of Bishop Bury is the true progenitor of the Bodleian. Balliol College, Oxford, too, as we have seen, was founded by a Durham man, John Baliol, the lord of Barnard Castle, some time between 1263 and 1268.

Two of our later bishops, Bishop Lightfoot and Bishop Westcott who followed him, have written many learned theological works. It was during Bishop Lightfoot's term, and with his hearty co-operation, that, in 1882, a new diocese was made of Northumberland. The bishopric has since then practically coincided with the county.

We have already spoken in the history section of the warriors of the Middle Ages, men who were repelling

the Scots, or again were in alliance with them.   But even
in modern days there have been Durham men who have

Sir Henry Havelock

fought with no less glory, and at the head of them must
come the gallant Henry Havelock, the Indian Mutiny

hero, who was born at Ford Hall near Sunderland in 1795, and died just after the relief of Lucknow.

Honour and fame come not only to great warriors and men of learning but not less worthily to the inventors and pioneers in commerce. It is to the "captains of industry" that Durham owes its modern importance. There were first of all the men connected with the pioneer railway. Edward Pease of Darlington, born in 1767, was the leading spirit, and with him were associated many other Quakers of that town. The first line was indeed nicknamed the Quakers' Line from the number of members of that body connected with it. The work of George Stephenson, whom Pease engaged in 1821 as engineer of the line, has already been noticed.

Another far-seeing capitalist and able organiser was Ralph Ward-Jackson, to whom it is due that a town of West Hartlepool now exists. Its growth dates from 1865, when the North-Eastern Railway took over the Hartlepool docks, and Jackson selected what was then waste sand as the site of the new port to be established. His name is remembered by the Jackson Dock and by the beautiful Ward-Jackson park. In like manner Sir Charles Palmer, who was born at South Shields, was the creator of the modern Jarrow. Here he established the works of the company of which he was founder and for many years chairman. The growth of Jarrow round the works— the blast furnaces, rolling mills, engine works, and ship-building yards—is a nineteenth century romance.

William Shield, born at Swalwell in 1748, the son of a boat-builder, began early to study music, became a conductor of orchestra, and ultimately Master of the King's Musicians.   William Booth, best known as "General" Booth, the founder of the Salvation Army, was born in Nottingham, but was for the years 1858–1861 minister of the New Connexion Chapel in Melbourne

Elizabeth Barrett Browning

Street at Gateshead, and was superintendent of the circuit in this town.

Joseph Ritson, the antiquary, critic, and collector of English literature, was born in Stockton in 1752 and educated there, but went to London while still a young man. Still better known in the world of letters was the poetess Elizabeth Barrett, who was born at Coxhoe Hall, near Durham, in 1806, and married Browning forty years later, dying in 1861. Her life was mostly spent away from her birthplace, in Herefordshire, London, and Florence.

John George Lambton, first Earl of Durham

She wrote several remarkable poems, the best-known, perhaps, being *Aurora Leigh* and the *Sonnets from the Portuguese*, but her *Cry of the Children* is of especial interest as a revelation of the wrongs of factory children.

The famous Lord Durham (John George Lambton,

first Earl of Durham), the originator of our self-governing colonies, also had much to do with the passing of the First Reform Bill in 1832. He was greatly in favour of the extension of the franchise, and was known as "Radical Jack." He was sent to Canada in 1840 as Governor-General to inquire into the disturbed state of the colony ; and the "Report" that he drew up is justly described as the most instructive document ever penned on colonial matters. The plans he suggested have now been adopted for all our great colonial possessions.

We have noted the invention of the safety lamp for mines and of the lifeboat as specially connected with our county. Another most useful, though slight, invention originated in Stockton. John Walker, a chemist in that town, was preparing some lighting composition when, by the accidental friction on the hearth of a match dipped into the mixture, ignition took place. This was the first "lucifer," the practical outcome of which was that in 1827 the inventor sold matches at the rate of 50 a shilling.

Sir Walter Scott spent a great deal of time in Durham. He admired highly the romantic scenery of the upper Tees, the "Rokeby Country" as it is called. Barnard Castle is the chief scene of the poem and Greta woods and Brignall banks are just over the border. At Old Park, near Bishop Auckland, the poet Gray used to visit his friend Dr Wharton.

25. THE CHIEF TOWNS AND VILLAGES OF DURHAM.

(The figures in brackets after each name state the population of the place according to the Census of 1911. Those at the end of each paragraph are references to the pages in the text.)

Barnard Castle (4757), a market town on the Tees. The magnificent Bowes Museum here contains a valuable collection of works of art. Its old castle is the chief scene in Sir Walter Scott's poem *Rokeby*. Greta Woods are quite near. (See pp. 27, 76, 138.)

Billingham (4463), an ancient Saxon village near Stockton, has a very old church (about 1260) in the Early English style. (p. 124.)

Birtley (8109), on the main North-Eastern line half-way between Durham and Newcastle. Mining is the chief industry, but clay and stone are quarried for building material.

Bishop Auckland (19,771), now the single residence of the Bishop. Auckland seems to show that oaks were once plentiful here: as Leland the antiquary says, "by the name it apperith to have been ful of okes." It is since the colliery development the centre of a populous mining district including Auckland St Andrew (5605), Coundon (6912), Escomb (2783), and West Auckland (4471). It has a fine situation on an eminence near the junction of the Gaunless and the Wear. (pp. 26, 127.)

Blaydon (9735), a rapidly-growing town on the Tyne, three miles above Gateshead. Coal-mining, iron-working, and brick-making are the industries: at Derwenthaugh new coaling staithes have not long been completed. Blaydon is on the North-Eastern line from Newcastle to Carlisle. (p. 67.)

Boldon (2982), not far from Sunderland, was the little village taken as type by Bishop Pudsey's inspectors who compiled the "Boldon Buke." It has a beautiful church of Early English architecture, with a curious tower and spire. (pp. 48, 108, 126, 131.)

Brancepeth, five miles south of Durham City, now a great colliery district, but famous for its castle and its church, which is curious for its mixture of styles. The tower is Early English, the nave and the transept are of the Decorated period, the chancel mainly Perpendicular. (pp. 26, 34, 38, 112, 133, 140.)

Brandon and Byshottles (17,667), a crowded mining district (chiefly coal, but also spar and barytes) three miles south-west of Durham. (p. 38.)

Castle Eden, a mining village encroaching on Castle Eden Dene round which colliery development is very active. (pp. 51, 135.)

Chester-le-Street (10,381), an ancient town on the Wear below Durham, is now the centre of a colliery district. The town itself is one of the very few where the population has decreased since the last census. Here was the mother church of the cathedral; it now has an Early English church with a lofty spire. (pp. 23, 26, 105, 116, 126, 130.)

Chilton (6070, in 1901 only 1411), a rapidly developing colliery settlement near Ferryhill.

Chopwell (10,029, the population in 1901 was only 4354), a rapidly developing mining and iron-working district. The Crown forest lands are here in the Derwent basin. (p. 72.)

Consett (11,207), on the North-Eastern loop between Durham and Newcastle. A vast iron and steel works occupies most of the inhabitants; coal mines also are worked; and there are iron mines, but they are unworked—foreign ore being brought more cheaply from Spain. (pp. 23, 24.)

Cornforth (5895), a closely-built colliery town in the Spennymoor district.

Crook (6630), a mining town four miles south-west of Durham.

Dalton-le-Dale, a small village about two miles from Easington, has a church, consisting of nave and chancel only, of late Norman origin.

Darlington (55,631), a municipal borough and a Parliamentary constituency returning one member. It is situated on the southern verge of the coal field, is the converging point of five railway lines, and is on the main line of the East Coast route. Recently the North-Eastern has transferred all its locomotive building works to the town. Iron and steel works and bridge-building employ most of the workers; the heaviest kind of goods appear to be more especially the staple product. It is built on the Skerne not far from its junction with the Tees. Its Early English church is, on account of its dimensions and architecture, of the first importance. (pp. 29, 77, 79, 80, 121, 127, 157, 158.)

Dinsdale Spa, about four miles from Darlington, pleasantly situated in the farming district that is so rapidly disappearing, is said to possess medicinal waters (containing sulphate of soda and muriate of soda) equal to those of Harrogate.

Dunston (9272), on the Tyne opposite the great Elswick Works where a number of its men are employed, is a rapidly growing place, and since the deepening of the river and the erection of coal staithes has become of importance for the shipping of coal. (pp. 4, 24, 86, 95, 145.)

Durham (17,550), the capital, a parliamentary constituency returning one member, the seat of the Northern University. Its own interest is merged in that of its cathedral, magnificently situated on a rock almost encircled by the Wear half-way between its source and its mouth. Durham School was founded by Henry VIII. (pp. 2, 8, 22, 26, 75, 136, 147, 152, 157, 158.)

Durham School

Easington (2711), a small, ancient village, now a busy colliery region, which gives its name to one of the five wards. It stands near the sea, two miles south of Seaham Harbour. It has a famous church. The nave and the chancel walls are Early English; but the beautiful windows of the chancel belong to the Decorated period. (pp. 51, 162.)

Ebchester (510), a mining village on the Derwent on the site of the Roman camp. (pp. 104, 146, 147.)

Escomb (2783), an ancient village on the Wear, 1½ miles above Bishop Auckland. Its small church is said to be Saxon dating from the seventh century: some of the stones are probably from the Roman station of Binchester, a short distance further up the river. (p. 124.)

Esh (10,175, only 7830 in 1901), a rapidly-growing mining town near Durham.

Felling (25,026), an extension of Gateshead towards the sea, and in reality one town with it.

Ferryhill (10,133, as compared with 3123 in 1901), a mining (coal, clay, lime) town, on the main North-Eastern line between Darlington and Durham. From Ferryhill are important loops of the railway, one west to Bishop Auckland, one east to the Hartlepools.

Frosterley, a mining and limestone-quarrying village in Weardale above Bishop Auckland. (p. 26.)

Fulwell, properly a residential suburb of Sunderland, famous for its raised beaches.

Gateshead (116,917), a county borough and a parliamentary constituency returning one member, the second town in the county. It was a Roman station, probably *Gabrosentum*, mentioned in the register of stations called the *Itinerary*; and the name may mean, as the arms of the town suggest, the "goat's head." It is now a busy industrial town with great engineering works, manufactures of chemicals, breweries, and brass-works. It is said that Daniel Defoe wrote the *Adventures of Robinson Crusoe* here, where he had sought refuge when "sorely prest by persecuting foes." The site of the Norman church is now occupied by one of the late Perpendicular period (about 1450). A doorway on the south side—with a peculiar zig-zag label—is the one remnant of the Norman building. (pp. 23, 24, 69, 79, 95, 132, 146, 156, 158.)

Greatham, two miles south of West Hartlepool, famous for its salt even in Edward the Confessor's time, now the site of the Cerebos Salt Company. Greatham Hospital represents one of the ancient charitable institutions of the bishops. (pp. 36, 89, 134.)

Hartlepool (20,615), also called East Hartlepool to distinguish it from its younger though larger neighbour, a municipal borough north of the mouth of the Tees. The naturally strong situation was at one time fortified, and some parts of the old walls and gateways are still standing. The fine church of St Hilda has a beautiful late Norman doorway. (pp. 37, 45, 52, 56, 86, 94, 95, 124, 127, 142, 157, 158.)

Haughton-le-Skerne (1359), an agricultural town on the Skerne, the tributary of the Tees, not far from Darlington. It has a church mainly of Early English construction but with the nave rebuilt in the Decorated style.

Hebburn (21,763), on the Tyne near Jarrow and in reality one town with it. Like the other Tyne towns it has shipbuilding and engine works, rope and sail factories, and is the home of many colliers who work in the district round. It stands half-way on the great water street that stretches from Gateshead to South Shields. (pp. 25, 69, 95.)

Hetton (15,678), a crowded mining town near Houghton in mid-Durham.

Houghton-le-Spring (9753), a coal-mining town at the head of a fine vale, six miles north-east of Durham. (pp. 126, 168.)

Hylton (3038), near the Wear mouth and really one with Sunderland. It has a fine castle dating from about 1260. (p. 141.)

Jarrow (33,726), a borough on the south bank of the Tyne half-way between Gateshead and the sea. It might almost be called Palmer's town for practically all the inhabitants are dependent on the immense shipbuilding and ironworks of that firm. The name of the town is the Saxon *gyrwy* (marsh or fen), and this refers to the extensive pool on the east side called Jarrow Slake. At one time it covered 500 acres but has lately been largely reclaimed. Jarrow is perhaps best known as the residence during the greater portion of his life of the Venerable Bede, the father of English learning. Next to the remains at Lanchester, portions of the Saxon monastery still remaining as part of St Paul's church are the earliest architectural relics in Durham. (pp. 25, 69, 76, 77, 95, 100, 118, 132, 145, 157, 164.)

Lanchester (5208), a coal-mining town on the railway loop from Durham to Consett. It stands near the site of the Roman camp and fortress, the stones of which served to build the ancient Norman church. The present church is mainly Early English (about 1250). (pp. 26, 104, 116, 147.)

Leadgate (4996), a colliery town in the vale of Derwent.

Marley Hill (2135), an almost isolated mining village on the hills above the Tyne, south of Gateshead. The highway from Gateshead leads through a charming agricultural and wooded district.

Medomsley (6221), a mining town beside the Derwent, two miles west of Chester-le-Street. Its church is in the Early English style. Its stunted spire and low flat-roofed nave are well suited to its exposed position. (p. 131.)

Middleton-in-Teesdale (1863), the terminus of the North-Eastern Railway up the Tees. The town might well still be called by its old name Middleton-one-Row. The district is agricultural; but there is mining of ironstone and spar, and quarries of building stone and whinstone. (pp. 27, 34, 151.)

Monkwearmouth, now included within the borough boundaries of Sunderland. There was a famous monastery here and portions of the Saxon building are yet to be seen. (pp. 118, 119, 152.)

Newbottle (7191), a mining town near Houghton-le-Spring.

Pelton (8118, as compared with 5504 in 1901). Colliery development has caused the increase in this town in the Durham City District.

Piercebridge (209), on the Tees between Darlington and Barnard Castle. The village is on the site of a Roman station where Watling Street crossed the Tees. The surrounding district is agricultural. (p. 147.)

Pittington (2130), three miles east of Durham, is a mining village. The tower and north side of the nave in the church are of Norman construction, the rest of the church mainly Early English (about 1260). (p. 127.)

Port Clarence stands at the mouth of Tees and is really the northern extension of Middlesborough (in Yorkshire). (pp. 27, 29.)

Raby, a little village that has grown up around the great castle of Raby, north of the Tees between Piercebridge and Barnard Castle. (p. 138.)

Roker (9965), part of Sunderland north of the mouth of the Wear, an ideal health and pleasure resort for the workers of Sunderland whence ready access is provided. (p. 48.)

Ryhope (11,185), a mining and iron-working town, two miles south of Sunderland. (p. 85.)

Ryton (12,948), on Tyne above the junction with the Derwent. The steepled church is of Early English style, probably about 1250, and the Rectory adjacent is in parts Elizabethan. The town is very pretty though the mines are close by. (pp. 126, 143.)

Sadberge (412), a little village between Darlington and Stockton, once the centre of a large district, the wapentake of Sadberge. At first exempt from the jurisdiction of the Palatinate, it was purchased from the impecunious Richard I, then planning his Crusade, by Bishop Hugh Pudsey. (pp. 108, 116.)

Seaham (6342), a coal-mining and lime-quarrying town south of Sunderland.

Seaham Harbour (15,757), a modern and rapidly-growing town (its population was 10,163 in 1901) due to the making of a secure harbour in the line of cliffs six miles south of Sunderland. (pp. 35, 49, 95.)

Seaton Carew (2265), a pleasant watering place and residential quarter south of West Hartlepool. (pp. 53, 95.)

Sedgefield (3327), an agricultural village on the North-Eastern between Ferryhill and Stockton. It is the site of the extensive County Lunatic Asylum. The quadrupled columns of the nave in the church with their fine foliated capitals are from about 1200. The rest of the church is more modern.

Shildon (13,488), two miles south-east of Bishop Auckland, an important mining centre on the borders of the agricultural district to the south.

Shotton (12,561, the population in 1901 was only 1917), a town near the coast six miles south of Sunderland. The town is an instance of extraordinary growth owing to the development of the collieries in its neighbourhood.

South Shields (108,647), a county borough, and a parliamentary constituency returning one member. It was most likely the Roman station *Ostia Vedra*; and was not deserted till the Romans finally left Britain, for a beautiful gold coin of Marcus Aurelius has been found here. The modern name is derived from the " sheelings " or sheds used by the fishermen; but it was originally known as St Hild's after the ancient chapel: even after the Restoration it is spoken of as " St Hild's, commonly called Shields." It has a fine situation at the mouth of Tyne, and so is admirably placed for its staple industries—shipbuilding, engineering, trading, and fishing. Steam ferries connect the town with Tynemouth and North Shields. (pp. 4, 25, 45, 47, 80, 95, 116, 156, 158.)

Southwick (13,784), adjacent to Sunderland on the north and forming one town with it.

Spennymoor (17,909), a rapidly-growing town six miles south of Durham. Besides coal-mining, it has an immense iron and steel works. (p. 159.)

Staindrop (1380), an ancient manor of the bishop near the Tees east of Barnard Castle. The nave of the church, the burial place of the Neville family, is of the transition period between Norman and Early English. The bulk of the work is Early English. The altar-tomb monuments of the Nevilles are the most elaborate in our county. (p. 130.)

Stanhope (5769), far up in Weardale and anciently the hunting headquarters of the bishop. The revenues of its church, derived from the lead-mines, were once over £12,000, one of the richest livings in the kingdom; but, since the almost complete cessation of the mines, they are now much diminished. Bishops Tunstall, Butler (who at Stanhope wrote the *Analogy of Religion*), and Phillpotts the "fighting bishop" of Exeter, were rectors of Stanhope before their elevation. (pp. 26, 34, 114.)

Stanley (population of the urban district is 23,294), a crowded mining and iron-working district on the south slopes of Tyne basin, west of Gateshead. (The district includes Annfield Plain, Burnopfield, Dipton, and Tanfield Lea.)

St John's Chapel, the largest town in the upper Wear Valley, and near the terminus of the western spur of the North-Eastern. (pp. 26, 82.)

Stockton (52,154), an ancient port and borough that has lately become an important modern industrial town and, since the deepening of the Tees, a deep-sea port. It gives its name to one of the five wards of the county, and returns one member. The Stockton and Darlington Railway of 1825 was the pioneer of our railway system. Its main industry is iron and steel-working, the iron mines of the Cleveland Hills and an abundance of good steam coal being easily accessible. It stands six miles above the mouth of Tees, and originally grew up around an ancient manor or castle of the bishop. (pp. 23, 29, 76, 95, 141, 148, 157, 167.)

Stranton, a picturesque little village with a beautiful church of the Decorated period. It stands near the sea, about a mile from Hartlepool.

Sunderland (151,159). The largest town in Durham, the twenty-second in Britain, a county borough and a parliamentary constituency returning two members. It is the commercial capital and the largest seaport of the county: its exports are mainly coal and coal products, steel rails, machinery of all kinds, fire-bricks, paper, and glass work; its imports are chiefly timber. Shipbuilding and marine engineering are its great industries: in a recent year one-sixth of the output of Britain came from Sunderland yards. Though a great industrial town it encloses a moor and has a seaside resort within a few minutes' ride. (pp. 9, 26, 45, 48, 58, 69, 79, 86, 95, 97, 100, 156, 158.)

Swalwell (3889), a mining and iron-working village near the Tyne three miles west of Gateshead. (p. 164.)

Tanfield (10,191), four miles south of Gateshead. It is a coal-mining, lime-quarrying and iron-working town. Ravensworth Castle stands near.

Tow Law (4324), six miles west of Durham. A town high up on the moors mining coal, clay, and gannister. (p. 82.)

Trimdon (5259), a colliery and iron-founding town in the Easington district.

Tudhoe (7872), a mining village near Spennymoor.

Tyne Dock (9628), a modern town, continuous with Hebburn and Jarrow, that has grown up around the newly-made docks once forming part of Jarrow Slake. (pp. 25, 69, 86, 95, 97.)

Ushaw, two miles south-west of Durham. It is a mining village, and is the seat of the Jesuit College now connected with Durham University. (p. 161.)

Usworth (7986), four miles west of Sunderland: coal mining and brick-making are the great industries.

Washington (7821), near the Wear between Chester and Sunderland. Besides coal-mining it has large chemical manufactures. (p. 80.)

West Hartlepool (63,923), a county borough, and together with Hartlepool a parliamentary constituency returning one member. It is a town of wonderfully rapid growth and dates only from the opening of the first dock in 1847. Its admirable position for commerce and industry, near to the sources of coal, iron, and lime, and well protected by a headland from the varying weathers of the North Sea, together with the enterprise of the North-Eastern Railway, has speedily made it the second port of the county. The deep-sea fishing industry in particular has been encouraged and facilitated; a huge fish quay along the deepened Victoria Dock of 17 acres, finished in 1911, allows trawlers quickly to land their fish and to take in stores without moving from the dock. It is the fourth timber port in the country and is first for mining timber. (pp. 52, 53, 57, 86, 90, 97, 100, 156, 164.)

Westoe, a residential suburb of South Shields, and a popular watering place south of the mouth of Tyne. (p. 47.)

Whickham (population of the urban district 18,332, of the ancient village 2036), beautifully situated on a bank above the Tyne, a mile west of Gateshead. It has a fine old church.

Whitburn (4406), really a northern suburb of Sunderland. It is a delightful watering place on the dry limestone rocks. Coal and lime are mined. (pp. 43, 47.)

Willington (5962), a mining town near Bishop Auckland.

Wingate (10,890), a mining and iron-working town in the Easington district.

Witton Gilbert (7098), on the outskirts of Durham, the site of a ruined castle.

Wolsingham (3414), an ancient bishop's manor in Weardale on the edge of the hunting forest, Stanhope Common, now a mining village. (p. 26.)

Wynyard, north of Darlington, an agricultural village near which is Wynyard Park and House, the seat of the Marquis of Londonderry.

DIAGRAMS   181

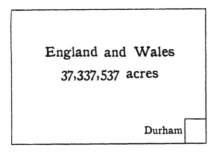

Fig. 1.   Area of Durham (649,244) acres) compared
with that of England and Wales

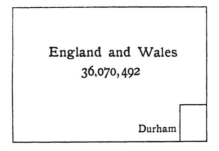

Fig. 2.   Population of Durham (1,369,860) compared with
that of England and Wales in 1911

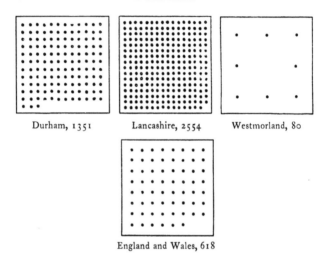

Durham, 1351   Lancashire, 2554   Westmorland, 80

England and Wales, 618

Fig. 3.   Comparative Density of Population to the
square mile in 1911

(*Each dot represents ten persons*)

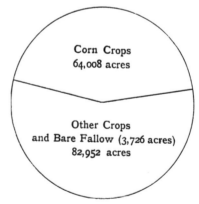

Corn Crops
64,008 acres

Other Crops
and Bare Fallow (3,726 acres)
82,952 acres

Fig. 4.   Proportionate area under Corn Crops compared
with that of other cultivated land in Durham in 1912

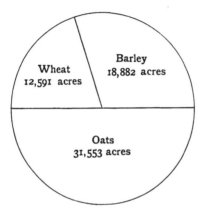

Fig. 5.   Proportionate area of chief Cereals in
Durham in 1912

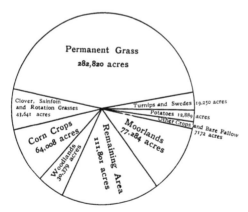

Fig. 6.   Proportionate areas of land in Durham in 1912

DURHAM

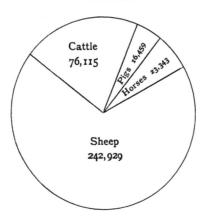

Fig. 7.   Proportionate numbers of Horses, Cattle, Sheep
and Pigs in Durham in 1912